THE WAY IT WORKS

Motion

PHILIP SAUVAIN

new Discovery BOOKS
NEW YORK

First American publication 1992 by New Discovery Books, Macmillan Publishing Company, 866 Third Avenue, New York, NY 10022

Macmillan Publishing Company is part of the Maxwell Communication Group of Companies

First published in 1991 by
Heinemann Children's Reference,
a division of Heinemann Educational Books Ltd,
Halley Court, Jordan Hill, Oxford OX2 8EJ

Library of Congress Cataloging-in-Publication Data
Sauvain, Philip Arthur
Motion / by Philip Sauvain
p. cm. — (The way it works)
Summary: Defines motion, describes its different types, and discusses how motion is used in bicycles, escalators, typewriters, and other types of machines.
ISBN 0-02-781077-1
1. Motion — Juvenile literature. 2. Force and energy — Juvenile literature. 3. Mechanical movements — Juvenile literature. [1. Motion 2. Force and energy. 3. Mechanical movements.] I. Title. II. Series
QC133.5.S28 1992
531'.1—dc20 91-24480

Photographic credits
t = top b = bottom r = right l = left

4 Allsport; 5 NHPA; 6 Trevor Hill; 9 British Waterways Board; 15 ZEFA; 17 British Coal; 19 Lego Group; 23 Anne Bolt; 25 Jones & Brother; 27 Allsport; 33 Don Morley/International Sports Photo Agency; 35*t* Robert Harding Picture Library; 35*c* Trevor Hill; 39 Science Photo Library; 41 Flymo Ltd; 43 Michael Blacker

Designed and produced by Pardoe Blacker Limited, Lingfield, Surrey, England
Artwork by Terry Burton, Tony Gibbons, Jane Pickering, Sebastian Quigley, Craig Warwick and Brian Watson
Printed in Spain by Mateu Cromo

91 92 93 94 95 10 9 8 7 6 5 4 3 2 1

Note to the reader

In this book there are some words in the text which are printed in **bold** type. This shows that the word is listed in the glossary on page 46. The glossary gives a brief explanation of words which may be new to you.

Contents

What is motion?

When you walk, run, or spin around, you are in motion. Even when you lie down, your chest moves up and down as you breathe. People are always in motion. Motion happens when anything moves.

Some motion is too slow for us to see. Sheets of ice called glaciers move so slowly down the mountains that we cannot see the ice move. A glacier may travel about five yards in a whole year. This is only about a hundredth of an inch an hour.

Some motion is too fast for us to see. The bullet from a rifle travels at a speed of over 500 yards a second. This is nearly two million yards an hour. A bullet moves four billion times faster than a glacier!

Making motion

Anything which moves has **energy**. Motion is a type of energy called **kinetic energy**. The other type of energy is **stored energy**. Even stationary objects, such as food and **gasoline**, have stored energy.

We need energy to move. We get energy from the food we eat. We use some of the stored energy in food for our motion. Motion always comes from a source of energy. A car engine uses the energy stored in gasoline to make the car move.

▼ Athletes have to train to be as fit and fast as this. The photograph has been taken with a high-speed camera.

Using motion

We use many different types of motion. When you pry open the lid of a can with a spoon handle, you use a form of motion. The spoon handle is acting as a **lever**. The lever puts a **force** under the lid to push it off. Pushing and pulling are ways of using motion. We can use force to press something down or to lift things up. We can also turn things around and around, like a wheel. This is called **rotary motion**.

Many of the tools and machines we see every day use these different types of motion. Some tools work like levers, such as a shovel or the hand brake on a car. Many machines, such as a typewriter, use up-and-down motion. When you press a key on a typewriter, it moves a letter to hit a piece of paper. Some of the tools in your home use rotary motion. A **food processor** and a rotary lawn mower use this type of motion to chop and cut. This book will show what some of these tools and machines do and how they work.

Snail's pace

Some animals move very slowly. A snail could take 10 hours to travel a quarter mile. When horses are racing they can travel the same distance in 40 seconds.

▶ *We use tools and machines to help us do different jobs. These machines and tools use different motions to do their work. Which motions are being used in these pictures?*

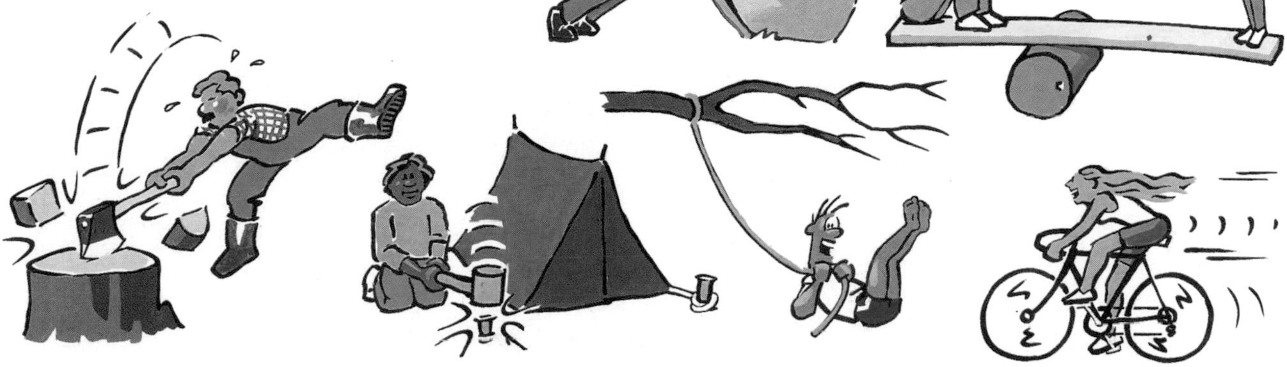

How the body moves

You can move your body in hundreds of ways. In fact, the human body is more amazing than most of the machines you will read about in this book. You can change direction instantly to move left, right, up, or down. You can start, stop, go slower, or go faster. You can jump, turn, push, pull, bend, kick, press, or sit.

People can make very small, careful movements. Doctors control their fingers to operate on tiny parts of the body, such as the eye. The human body can also do many different jobs. One moment a person may use skill to push cotton thread through the eye of a needle. The next moment, the same person may use strength to hammer in a tent peg or chop down a tree with an ax.

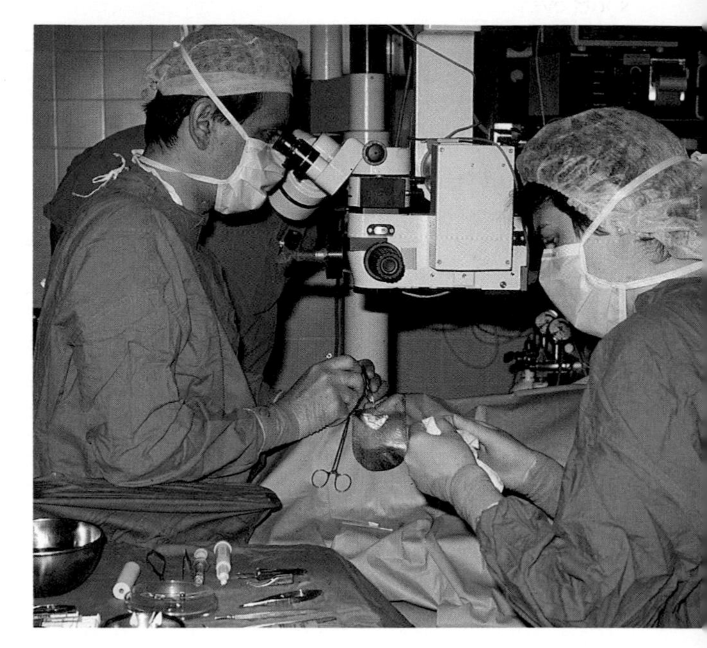

▲ People who do delicate jobs with their hands need steady fingers. They have to move with great care. A wrong movement could spoil their work.

Moving parts of the body

You move your arms and legs at the **joints**, such as the elbow, wrist, knee, and ankle. You use **muscles** to move a limb at the joint. With your shoulder muscles, you can lift your arm up high, bring it down, or swing it around.

There are muscles all over your body. Your brain sends messages to tell your muscles what to do. The brain is your computer. It makes all the parts of your body work properly. You can see many of the body's moving parts in this picture of the human body.

The heart has very strong muscles that pump blood to all parts of the body. The stomach helps us to release the energy stored in the food we eat. The stomach muscles also help us to breathe. When you cough you can feel your stomach muscles move.

Controlling movements

We use our muscles all the time. Often, we move many different parts of our body at the same time. Think what happens when you lift a spoonful of food to your mouth. You move your head and eyes to find your food. You move your shoulder, elbow, wrist, hand, and fingers when you lift the spoon. You move your mouth, jaws, teeth, and tongue to eat the food. Your chest muscles help you to breathe while you eat. Different parts of your body move the food through tubes to your stomach.

The brain controls all these movements. This is called **coordination**. It makes sure that all our movements work together in the right order and at the right time.

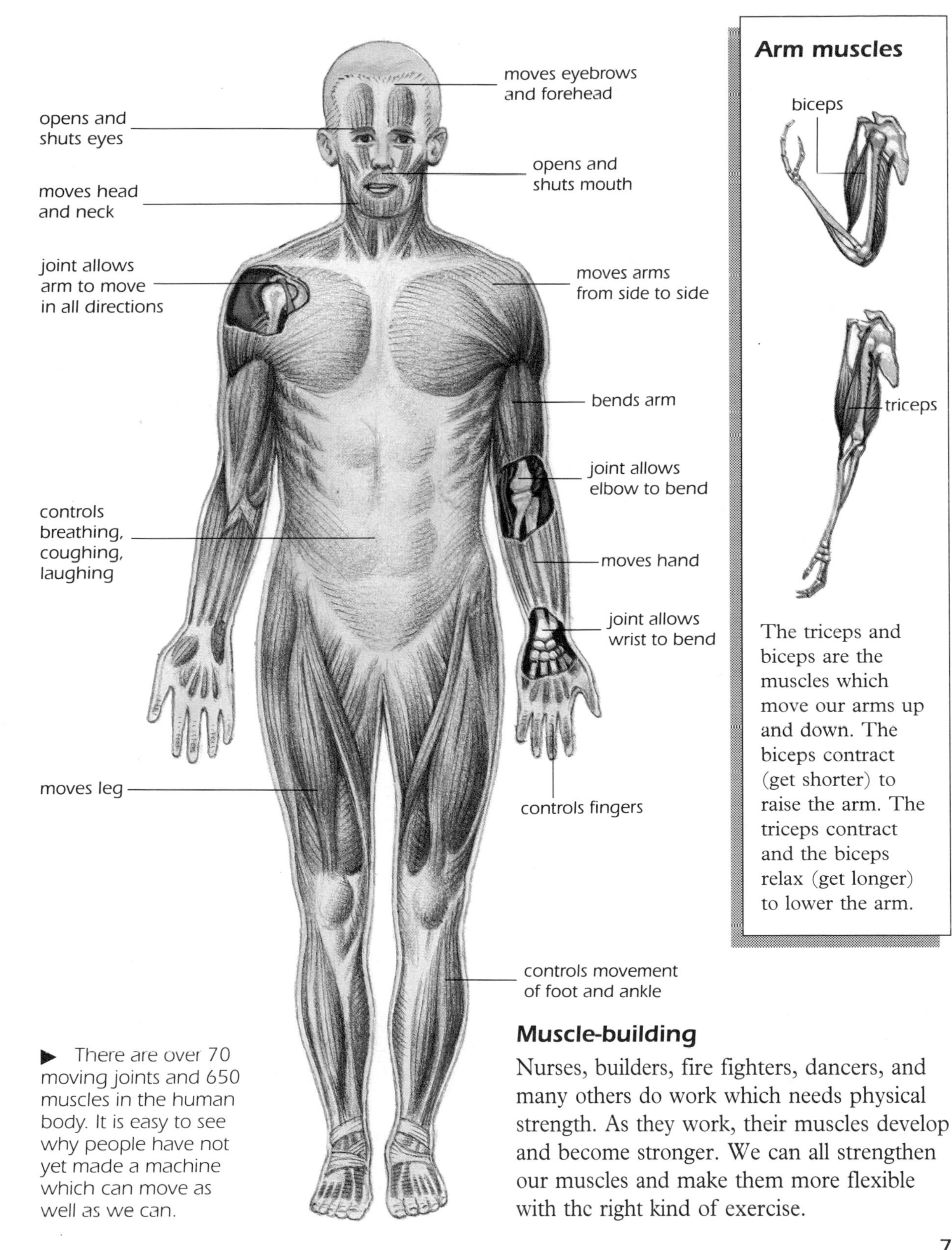

Arm muscles

The triceps and biceps are the muscles which move our arms up and down. The biceps contract (get shorter) to raise the arm. The triceps contract and the biceps relax (get longer) to lower the arm.

▶ There are over 70 moving joints and 650 muscles in the human body. It is easy to see why people have not yet made a machine which can move as well as we can.

Muscle-building

Nurses, builders, fire fighters, dancers, and many others do work which needs physical strength. As they work, their muscles develop and become stronger. We can all strengthen our muscles and make them more flexible with the right kind of exercise.

Using levers

The lever was one of the first tools. With a lever, people can move heavy objects using only a small amount of force. The simplest lever is a stick or pole. You put one end of the stick under the object you want to move. You push a stone under the stick close to the object. The point where the stick turns on the stone is called the **fulcrum**. When you press down on the top end of the stick, it forces the bottom end up. This lifts the object. The force used to push the lever is called the **effort**. The **load** is the object moved or lifted by the lever.

Simple levers

Many of the tools we use each day are simple levers, such as a shovel and a pair of scissors. A seesaw is another simple lever. The fulcrum of the seesaw is the point in the middle on which the plank turns. The child at one end is the load. The child on the other end is the effort. When a heavy child sits down at one end, the seesaw lifts the lighter child at the other end.

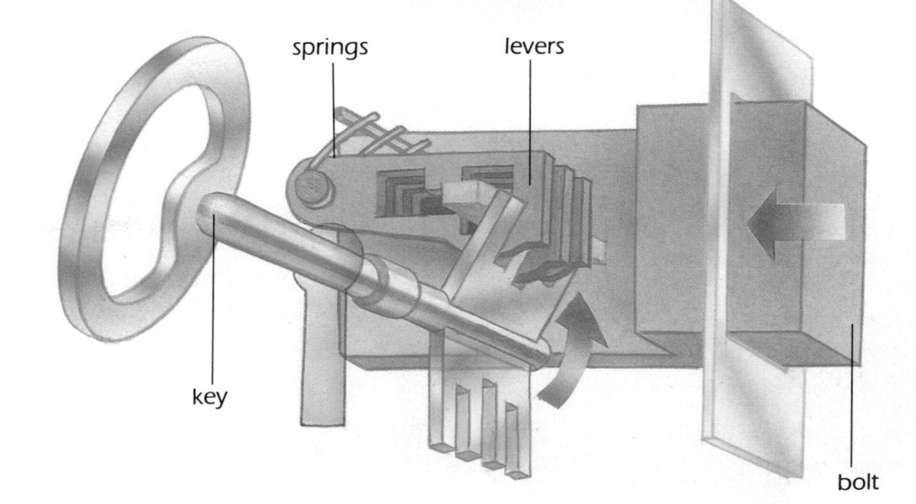

▶ *We use levers when we turn a key to lock a door. A bar moves across from the lock to a slot in the door. The key makes metal levers hold the bar firmly inside the slot. We need a key shaped to fit the levers to release the bar and open the lock.*

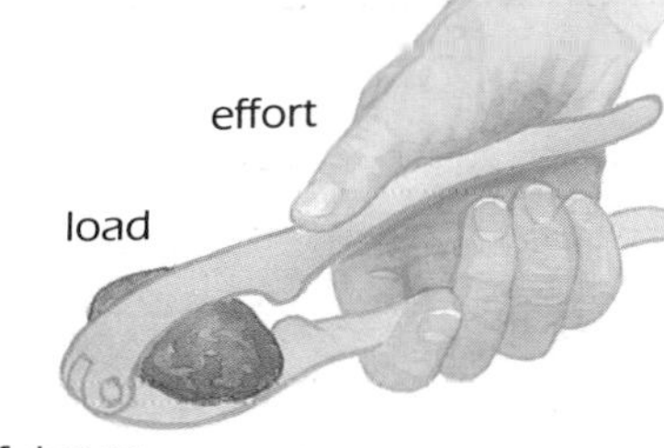

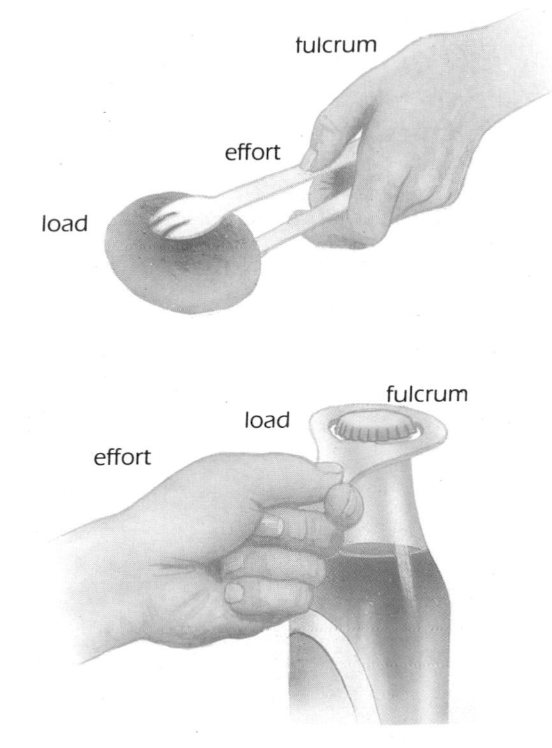

◄ There are three main types of levers. First-order levers have the fulcrum in the middle. Second-order levers have the load in the middle. Third-order levers have the effort in the middle.

Complicated levers

Simple levers, such as a stick, shovel, and seesaw, are **first-order levers**. They have the fulcrum somewhere between the load and the effort. For example, the fulcrum may be in the middle, with the load at one end and the effort at the other.

More complicated levers have the fulcrum at one end, instead of somewhere in the middle. You apply effort to the other end of the lever. It is the load that is somewhere in the middle. A wheelbarrow is a lever like this. It is a **second-order lever**. The fulcrum is over the wheel. The load is in the barrow. You lift this load when you use effort to raise the handles at the other end. Nutcrackers and bottle openers are other second order levers which you may have at home.

With a **third-order lever** you use effort somewhere between the fulcrum, at one end, and the load at the other end. A pair of tongs is a third order lever. You squeeze the tongs in the middle to grip something at the end.

◄ A lock gate is a first-order lever. The lever is much longer than the gate, so one person is able to move the gate even if it is very heavy. Huge pieces of wood have to be used to make the gate, as the levers need to be very strong.

Measuring weights

We use a tool like a seesaw when we weigh things on a pair of **weighing scales**. When the weights in the pan on one side are equal to the weight of the goods in the other pan, the pans are level (at equal heights). We say that the weights **balance**. If you put extra weight in one pan, it sinks. To make the pans balance again you must add extra weight to the other pan.

Balancing things helps us to find their weight. This is why we call a pair of very accurate scales a balance.

In the balance

A seesaw balances when the weight of the child on one end is equal to the weight of the child on the other end. When a heavy child sits at one end, it sinks. The seesaw lifts the lighter child at the other end into the air.

If the heavy child sits on the bar of the seesaw, nearer to the fulcrum, something strange happens. Instead of lifting the lighter child, the seesaw balances. This is because a weight on the seesaw pushes down with more force the farther the weight is from the fulcrum. So the lighter child farther from the fulcrum balances the heavier child nearer to the fulcrum.

If the heavier child moves even nearer to the fulcrum, he or she is lifted into the air. The weight of the lighter child on the longer end of the seesaw can now lift the heavier child on the shorter end. If one end of the seesaw was long enough, a small baby could lift the heaviest man in the world!

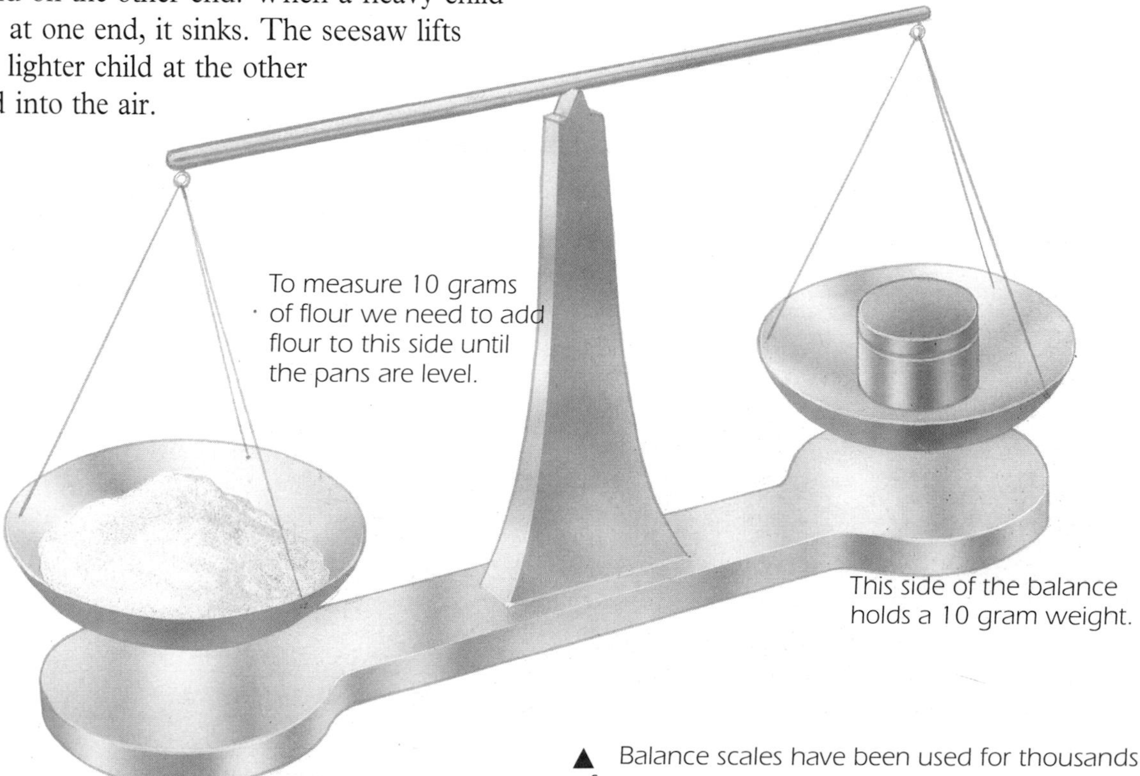

To measure 10 grams of flour we need to add flour to this side until the pans are level.

This side of the balance holds a 10 gram weight.

▲ Balance scales have been used for thousands of years.

pan holding substance to be weighed

spring pushed down by the weight of the goods in the pan

pointer linked to the spring

dial recording weight

▲ A seesaw is a type of lever. It is also a balance. A seesaw does not work well unless the children are about the same weight. If a heavy child pushes the end of the seesaw down, the light child will be stuck up in the air. If the heavy child moves closer toward the center, the seesaw balances.

▶ Kitchen scales use a spring and a dial to weigh food. The food presses down on the spring. A series of levers makes a pointer move around the dial. Bathroom scales also work like this.

Supermarket scales

The scales at a supermarket are electronic. When the items are put on the scale, an electronic device measures the movement in the scale and displays the weight on a small screen. Supermarket scales often include a calculator which works out the price of the goods. Some electronic scales can measure very small movements, so they can measure tiny weights accurately.

On a knife's edge

Nurses often use another type of **scale** to weigh people in a clinic or hospital. The person stands on a weighing platform which looks as if it rests on the ground. In fact, the platform hangs down from a steel beam balanced on a knife edge. The weight of the person pulls down the shorter arm of the beam. The nurse slides a heavy weight along the longer arm of the beam until the beam balances. A scale marked on the longer arm shows the weight of the person standing on the platform.

Lifting

A small child can lift a heavy weight using a lever, such as a seesaw. There are other kinds of machines we can use to help us lift weights. Like levers, they multiply the effort we use so that we can move heavier weights than we could if we were lifting them directly. The actions of pulling down on a rope or turning a wheel are easier than lifting up a heavy weight.

Lifting water

For thousands of years, farmers in Egypt have used water from the Nile River to water their crops. Lifting water is hard work. A simple machine called a shadoof has helped them to do the job for most of that time.

A long pole is like the beam of a scale. A heavy stone on the short end of the pole balances the weight of the bucket of water. We call the stone a **counterweight**. This makes the shadoof easy to swing around to empty the bucket.

Water wells

Many people take their water from deep holes in the ground, called wells. A person lowers a bucket on the end of a long rope down the well. When the bucket is full of water, the person pulls the rope up the well. A full bucket is heavy to pull up. Some wells are very deep, so people usually wind the rope around a wheel or pole at the top. A handle is fitted onto the wheel or pole.

▼ The shadoof is like a long seesaw. It acts like a lever to lift water from the river. The heavy stones at one end balance the bucket of water at the other end.

▶ The more pulley wheels that are used the easier the job of lifting becomes. Using two pulley wheels, you can lift an object twice as heavy as you can with one pulley wheel. Three pulley wheels let you lift a load three times as heavy.

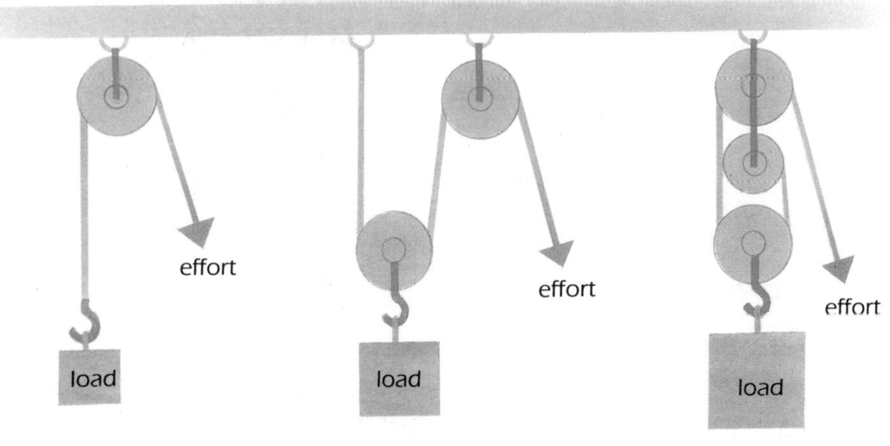

Lifting with wheels

Lifting heavy objects became easier when someone invented the first **pulley**. A pulley is a wheel with a groove around the rim. You tie one end of a rope to the object you want to lift. The other end of the rope passes over the groove in the pulley wheel. When you pull down on the other end of the rope, the weight rises. The pulley wheel changes the direction of the effort you apply. Instead of pulling up, you pull down. This makes it easier to lift a heavy object. You can use your own weight as a counterweight to help lift the object on the other end.

◀ To lift very heavy objects people use pulleys. Using pulley wheels, a heavy engine can be easily lifted by one person.

A crane

Today, we often use cranes to lift or move heavy objects, especially on building sites. Cranes like **jib cranes** and mobile cranes use counterweights in a similar way to the shadoof. The counterweight balances the load and stops the crane from falling over when it lifts a heavy load.

There are two main types of cranes. The jib crane is the type of crane you often see at building sites. It can be moved from site to site in pieces. The **bridge crane** is often built into a factory.

A jib crane

The jib crane can have a tall metal tower, like the upright poles on the shadoof. On one side of the tower is a long arm called the jib. The main arms of the jib are made of strips of steel with crossbars for support. So the jib is strong without being too heavy. Balancing the jib on the other side is a shorter arm and the counterweight. The driver can move the jib of the crane easily to lift heavy objects, such as a steel **girder**.

A mobile crane

You have probably seen a mobile crane like this going along the street. It is lifted off its wheels while it is working. The long arm is called a boom. The boom can be extended like a telescope and can swivel round. The counterweight at the base keeps it stable and balances the weight of the load.

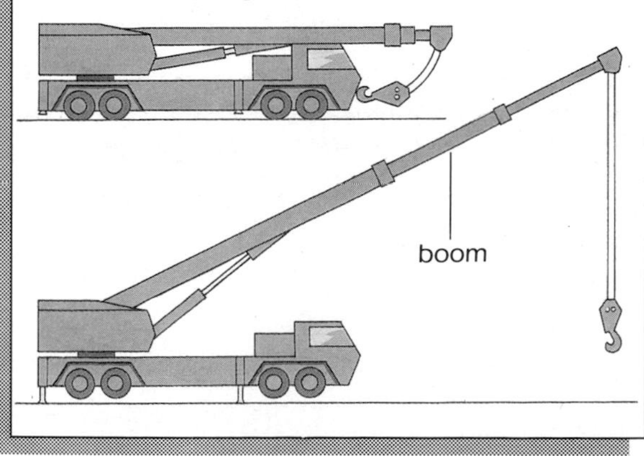

jib
pulleys
cables
load

To lift a number of objects at the same time, the driver uses a net instead of a hook. If the material is loose, like sand, the driver scoops it up with a **grab** or bucket.

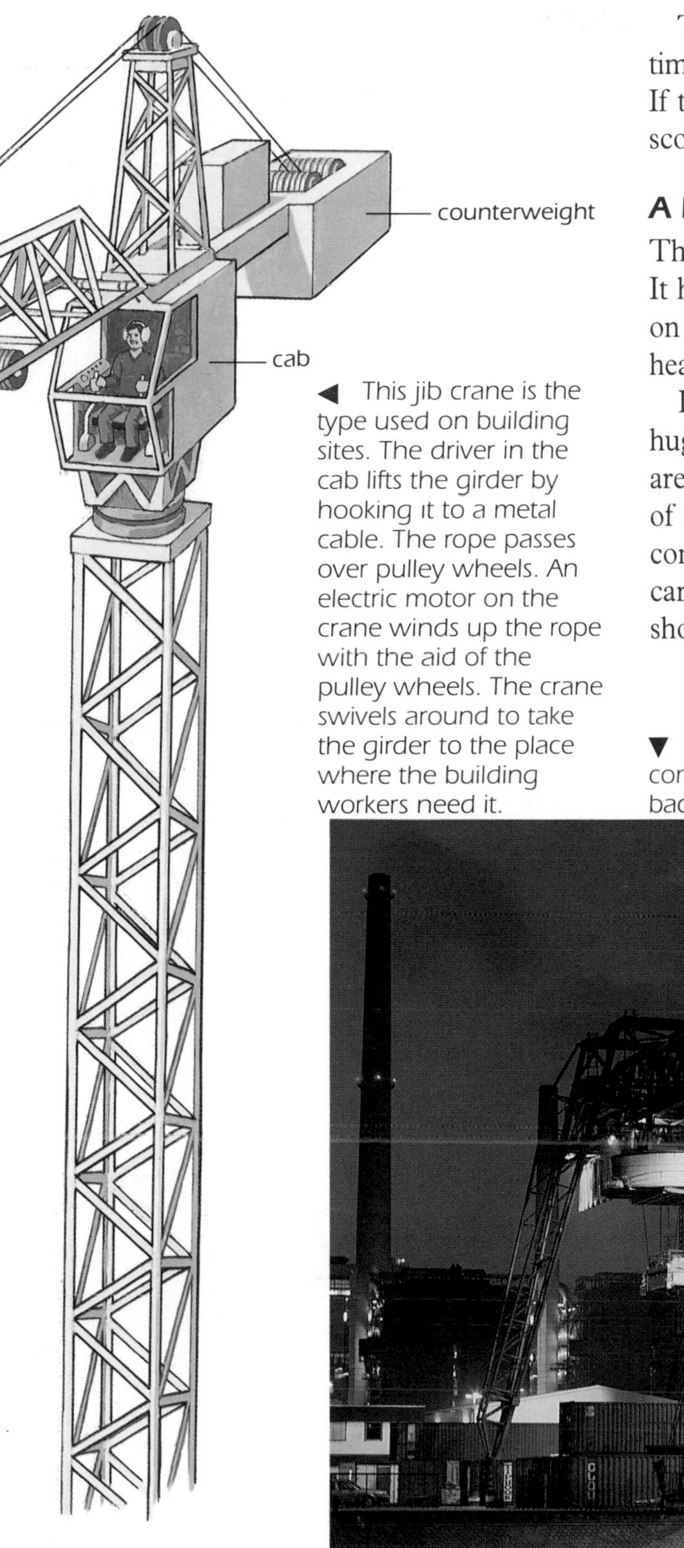

◀ This jib crane is the type used on building sites. The driver in the cab lifts the girder by hooking it to a metal cable. The rope passes over pulley wheels. An electric motor on the crane winds up the rope with the aid of the pulley wheels. The crane swivels around to take the girder to the place where the building workers need it.

A bridge crane

The bridge crane looks like a moving bridge. It has two legs and can move back or forth on rails or on wheels. Bridge cranes often lift heavy loads in factories.

Ports use bridge cranes to load ships with huge metal boxes called containers. These are packed with goods at the factory instead of at the docks. Cranes can unload and load containers quickly. In this way, a ship carrying containers needs to spend only a short time in the port.

▼ This bridge crane is used to load and unload containers at these docks. The crane moves backwards and forwards on rails.

The elevator

Some of the buildings we have today are very tall. People would not like to live and work in them if they had to use stairs to get to the top floors.

The first passenger elevator was used in a china shop in New York in 1857. A **steam engine** did the work of raising and lowering the elevator.

Today, **electric motors** usually provide the **power** for elevators. The World Trade Center in New York is 110 stories high. The elevators which take people to the top of the building get their energy from electricity.

How an elevator works

The part of the elevator in which you travel is called the car. Upright rails guide the car up and down the elevator **shaft**. The car is held by steel cables which pass over pulleys at the top of the building.

An electric motor near the roof provides the power to move the elevator. There is also a counterweight that is as heavy as the elevator car. As the elevator rises, the counterweight sinks down the elevator shaft on the other side. This counterweight does much of the work needed to pull the elevator up the shaft. The pulleys make the work of raising the car easier. The electric motor does the rest.

The speed at which the elevator car moves changes **automatically**, just before the elevator stops at a floor. This is why the elevator car moves slowly in the last few

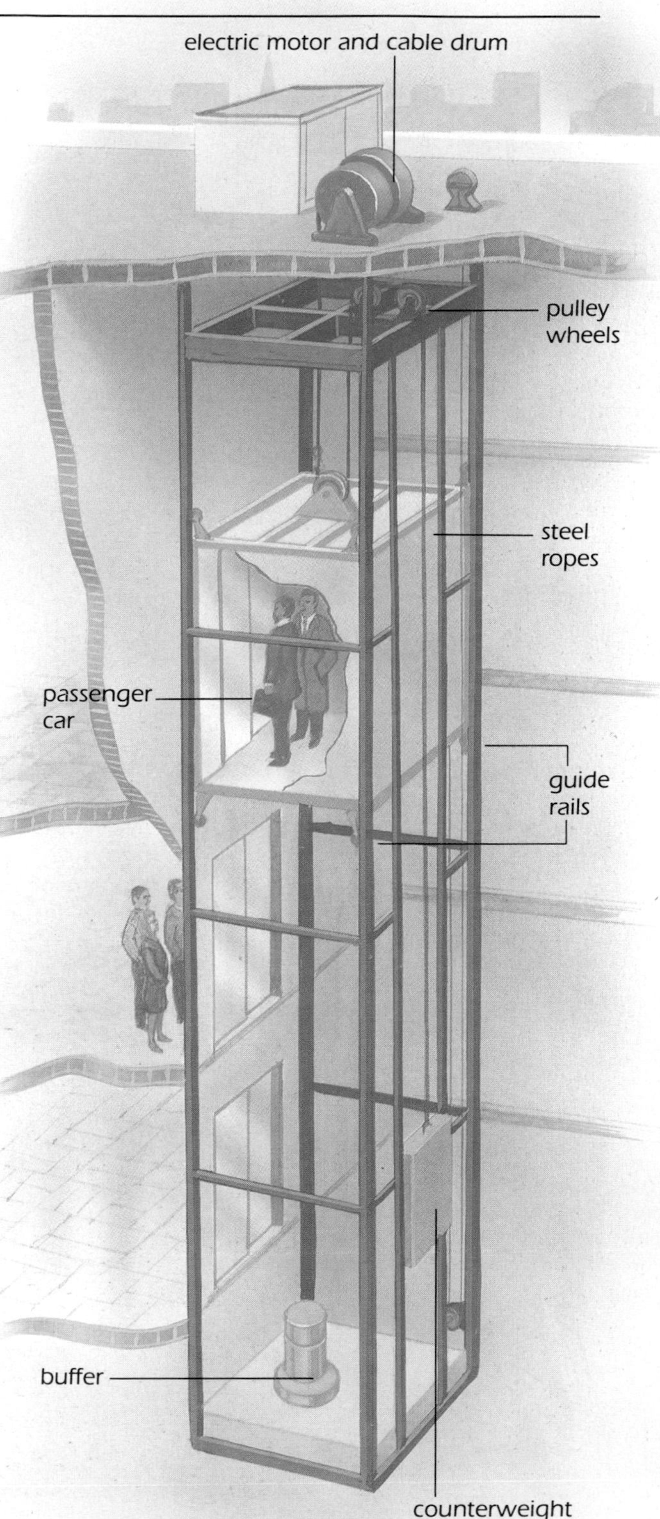

▶ *This is how an elevator in a tall building works. The lifting cables are made of strong steel. There is a small motor on the roof of the car to open and shut the doors.*

seconds before stopping. Next to the motor is a **governor**. This is a device which helps to control the speed of a machine. The governor stops the car from dropping too quickly to the ground. The buffer at the bottom absorbs the force of the landing car. A modern elevator has other safety devices to help prevent accidents.

▼ A mine elevator is different from a passenger elevator. The cage often travels much faster and much farther. The fastest mine elevator in the world travels down a shaft over 6,500 feet (2,000 meters) deep at a speed of 40.7 miles (65.8 kilometers) per hour.

In a mine

Many mines are very deep under the ground. The miners use elevators to go to work. The miners call the elevator car a **cage**. This may be because the car of a mine elevator often looks like a large birdcage. Some cages have two floors, so that two groups of miners can travel down the shaft at the same time. Mine elevators need to be fast. They often have to travel much farther than an elevator in a building. The world's tallest skyscraper is about three-tenths of a mile (half a kilometer) high. Some mine shafts go down six-tenths of a mile (one kilometer) or more below the surface.

Moving stairs

An elevator can only carry a few people at a time. More people can travel on a moving staircase called an **escalator**. It can move about ten times as many people in a day as an elevator car.

About 100 years ago, people in the United States used the world's first escalator. It was invented by Jesse W. Reno in 1894. He built his escalator for use as an amusement park ride at Coney Island, New York. Within five years there were escalators in large stores in New York and London.

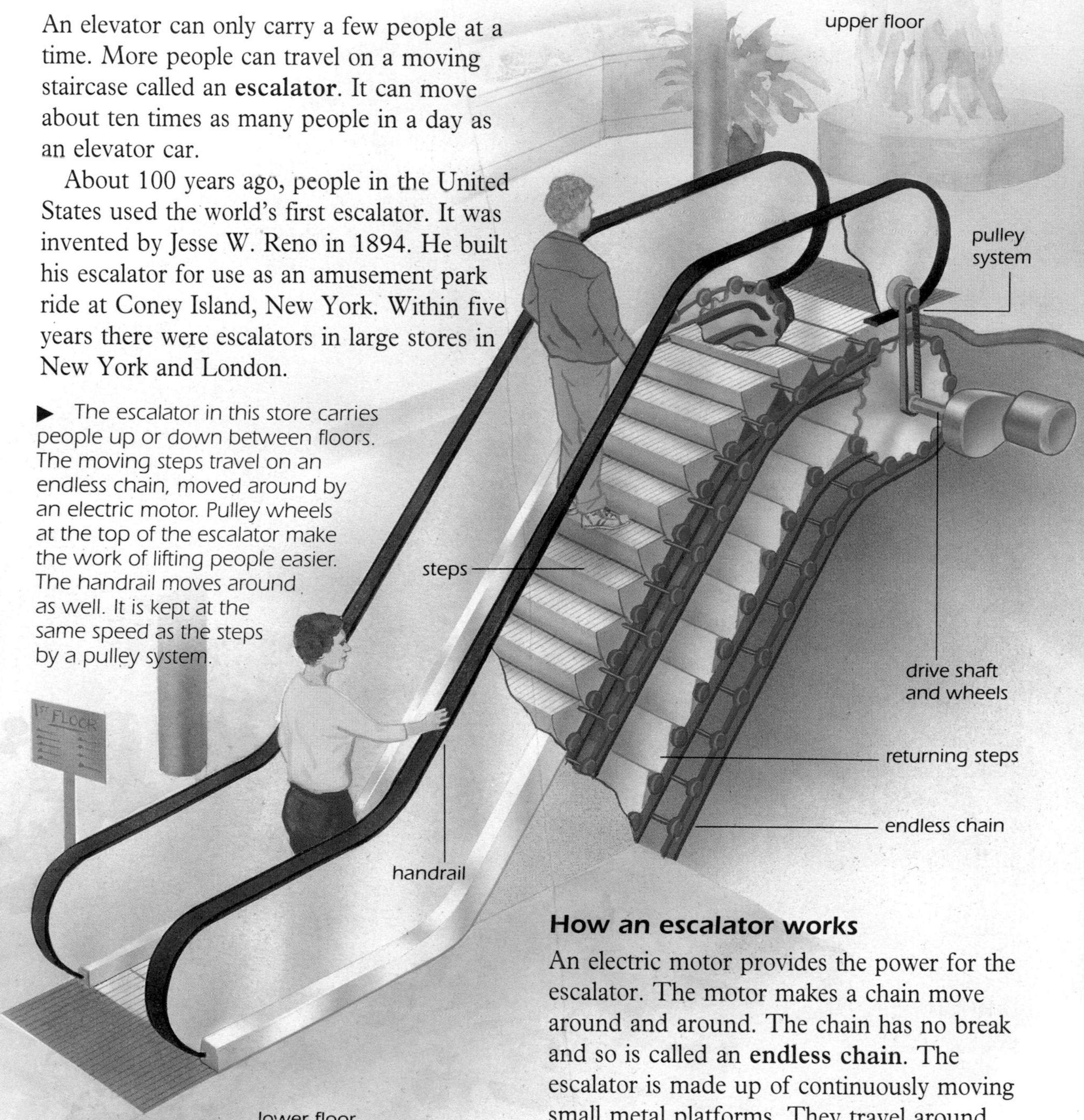

▶ The escalator in this store carries people up or down between floors. The moving steps travel on an endless chain, moved around by an electric motor. Pulley wheels at the top of the escalator make the work of lifting people easier. The handrail moves around as well. It is kept at the same speed as the steps by a pulley system.

How an escalator works

An electric motor provides the power for the escalator. The motor makes a chain move around and around. The chain has no break and so is called an **endless chain**. The escalator is made up of continuously moving small metal platforms. They travel around and around on the endless chain. Each platform has tiny wheels which move along on rails beneath it. As the escalator rises, the wheels slide along the rail to push each platform into a step. The rails hold the steps one above the other, like stairs. At the top of the stairs the rails and the steps flatten out. They turn downward with the chain and return, upside down, to the bottom. The steps are still flat when they appear again as the entrance to the escalator. The flattened steps make it easier for people to get on and off the escalator.

The handrail which you hold onto also goes around and around. The motor that drives the endless chain also moves the handrail around. The handrail curves around at the top and runs down under the floor to return again at the bottom of the stairs.

Other endless chains

Factories and mines often use a type of endless belt called a **conveyor belt**. It can carry goods from one end of a factory to the other. In some factories, parts are put together to make goods. The conveyor belt carries the partly made goods along the **production line**. As the items go through different departments, the factory workers add more parts to them.

Most coal mines use conveyor belts. They may even bring coal to the surface from the mine deep underground. The conveyor belt moves along a steeply sloping tunnel instead of up the mine shaft.

Moving along

Another kind of endless chain carries people along the ground. This is called a moving walkway. At airports, these walkways carry people and their bags to the aircraft.

▼ This is part of a factory production line. The items pass along a conveyor belt as they are made. The belt does not stop, so the people in the factory always have work to do.

Swinging motion

Keeping an escalator in motion is hard work. If the motor stops, the escalator stops. Keeping a swing in motion is much easier. You do not need to use force all the time to keep a swing moving. If you push the swing, it will move back and forth many times.

The swing is a type of **pendulum**. The amount of swing on one side of a pendulum balances the amount of swing on the other. The farther you pull the pendulum back, the more it swings forward. If you only pull the pendulum back a short way, it will only go forward a short way.

Keeping time

There is another unusual thing about the swinging motion of a pendulum. It takes almost the same time to move back and forth whether the swings are high or low.

You can test this for yourself. Make a pendulum by tying a weight to a long piece of string. Get a friend to hold the top of the string in one hand and the weight in the other. Ask your friend to pull the weight back as far as possible and then let it go. Start timing the pendulum as soon as the weight starts to move. You could use a watch or count as fast as you can. Stop timing when the pendulum reaches the end of its swing on the other side.

Now test the pendulum again, but ask your friend to pull the weight back only a short way. Again, time how long it takes for the weight to reach the other side. See how slowly the pendulum moves. Did this small swing take about the same amount of time as the big swing?

◀ When you are on a swing you are on a type of pendulum. If you swing low, you move slowly. If you swing high, you move faster. No matter how high you go the time taken to swing back and forth stays almost the same.

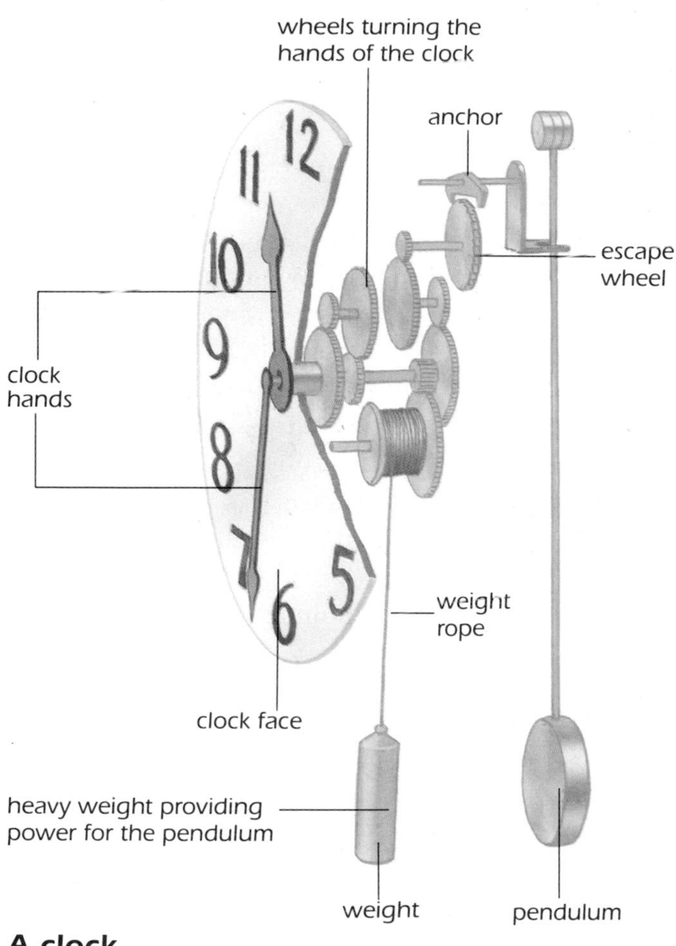

A clock

Since the swing of a pendulum always takes the same time, a pendulum can be used in a clock to measure time. At the top of the pendulum in such a clock is an **anchor** with two hooks at each end. As the pendulum swings, the anchor rocks. At the end of each swing, a hook on the anchor catches hold of a tooth, or **cog**, on the edge of a wheel. This is called the escape wheel and it is pulled around by weights. The anchor controls the time that the wheel takes to turn. While the pendulum swings back, the anchor tips up to free the wheel. The wheel then moves around a cog. Then the other end of the anchor catches the wheel again. This is why the clock ticks. Each tick is the anchor catching a cog.

▲ The pendulum makes sure that this clock always tells the right time. The escape wheel moves around because a heavy weight is pulling on the wheel. The pull of the falling weight also helps to keep the pendulum swinging. The swing of the pendulum keeps the escape wheel turning at a steady rate. The escape wheel turns other wheels which move the hands.

The swing of the pendulum makes the escape wheel turn at a steady rate. The cogs on the escape wheel fit between cogs on other wheels, called **gear wheels**. As the escape wheel moves around it turns the gear wheels, which turn the hands on the clock.

Striking a tune

A pendulum always swings the same distance to one side as it does to the other side. You will see a similar motion if you pluck a guitar string and look closely. The string bends rapidly back and forth. We call this type of motion **vibration**.

A guitar string makes a sound when you pluck it. Most sounds come from vibrations. The sound as you strike an empty can comes from the metal sides vibrating. You can produce a different sound by hitting a drum and making the skin vibrate. When you listen to a piece of music, you are hearing many different vibrations.

How a piano works

The sound of a piano is made by vibrating strings. Each piano key is linked by levers and rods to a felt hammer. When you press a

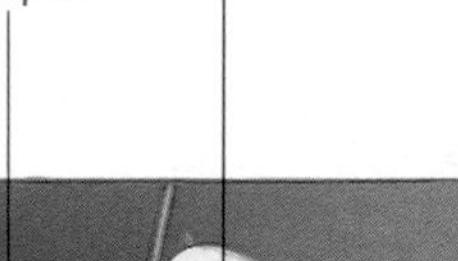

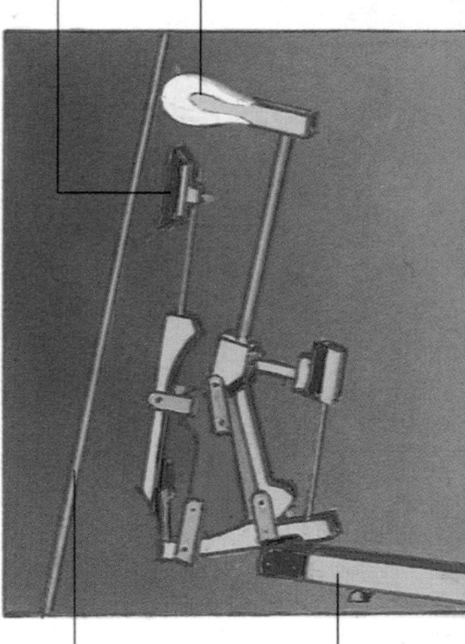

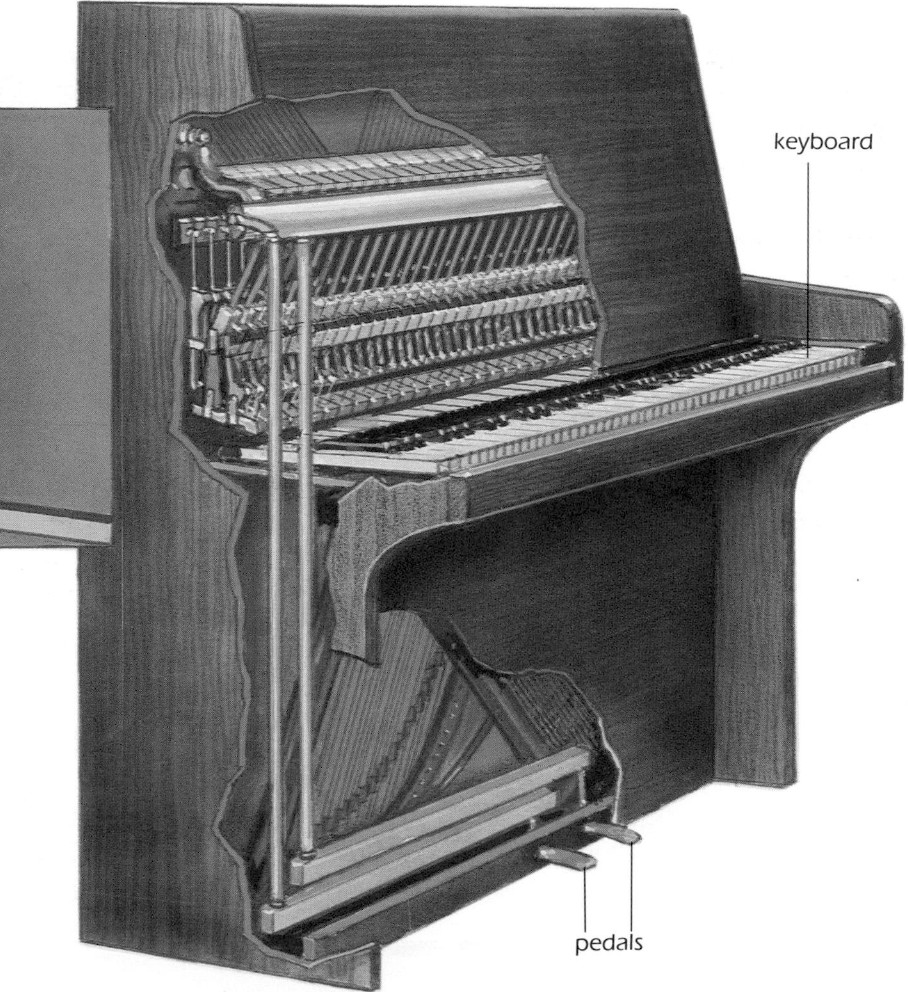

▶ Inside the piano is a complex system of strings, hammers, and levers. Pressing a key makes the hammer strike the string. The string vibrates and makes a sound. The damper then stops the vibration of the string.

key, you make the hammer rise. The hammer strikes a piano string. The string vibrates and we hear this as a note on the piano.

Each string vibrates to give a particular sound. The short, thin strings in the right half of the piano vibrate quickly. We hear these as high sounds. The longer, thicker strings in the left half of the piano vibrate more slowly, producing lower sounds.

A note on the piano stops as soon as you release the key. This is because a lever moves a **damper** against the string. The damper stops the string vibrating. Without the damper, the string would continue to vibrate for a long time. If you want the note to continue, you press a pedal under the piano with your foot. This pedal uses levers to move the dampers away from the strings. Pressing another pedal under the piano moves the hammers slightly. This makes the notes sound softer than normal.

▲ This is a steel band. Their instruments are made from old oil drums. The surfaces of the drums are curved metal pans. Each pan has several panels which make different sounds when they are hit.

Striking instruments

These are all percussion instruments. They make their sound when they are struck with sticks or hammers. They may be made of animal skin, wood, or metal. In an orchestra, the **percussionist** may play several of these instruments.

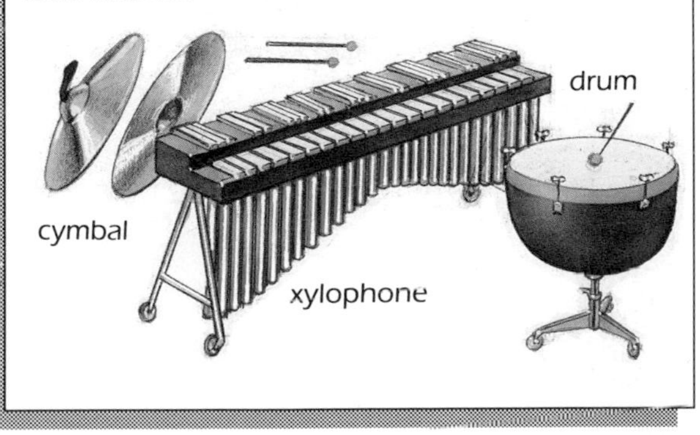

Playing a drum

A drum is a skin stretched tightly over a bowl or frame. Some drums have skins across the top and bottom. When the drummer hits the skin, it vibrates. The air inside the drum vibrates as well. These vibrations are the sounds we hear from the drum.

The drummer in a band usually plays a set of drums. A small drum gives a higher sound than a large drum. The sound also depends on how hard or fast the drummer strikes the drum. The skin vibrates in a different way when it is hit with the hands than when it is hit with a stick.

Striking a letter

We press keys to make a typewriter work, just as we play a piano. We provide the energy to make these machines work. As with the piano, the typewriter keys are levers. When you press one end down, the other end moves up and causes other levers to move in turn. The typewriter keys print letters on paper instead of striking strings.

How a typewriter works

Each of the typewriter keys is linked by rods and levers to a **type bar.** This has a letter, number or sign on the end. When you tap a key, the type bar lifts up to press an inked ribbon against the paper. This action marks the letter on the paper. The paper is wound around a roller called the **platen**. This is part of the moving **carriage** of the typewriter.

Each time you type a letter, a spring moves the carriage to the left. At the end of the line, you pull a lever to move the carriage up and across. The lever also makes the platen roll around so a clean section of paper is in position. Then you can start typing the next line.

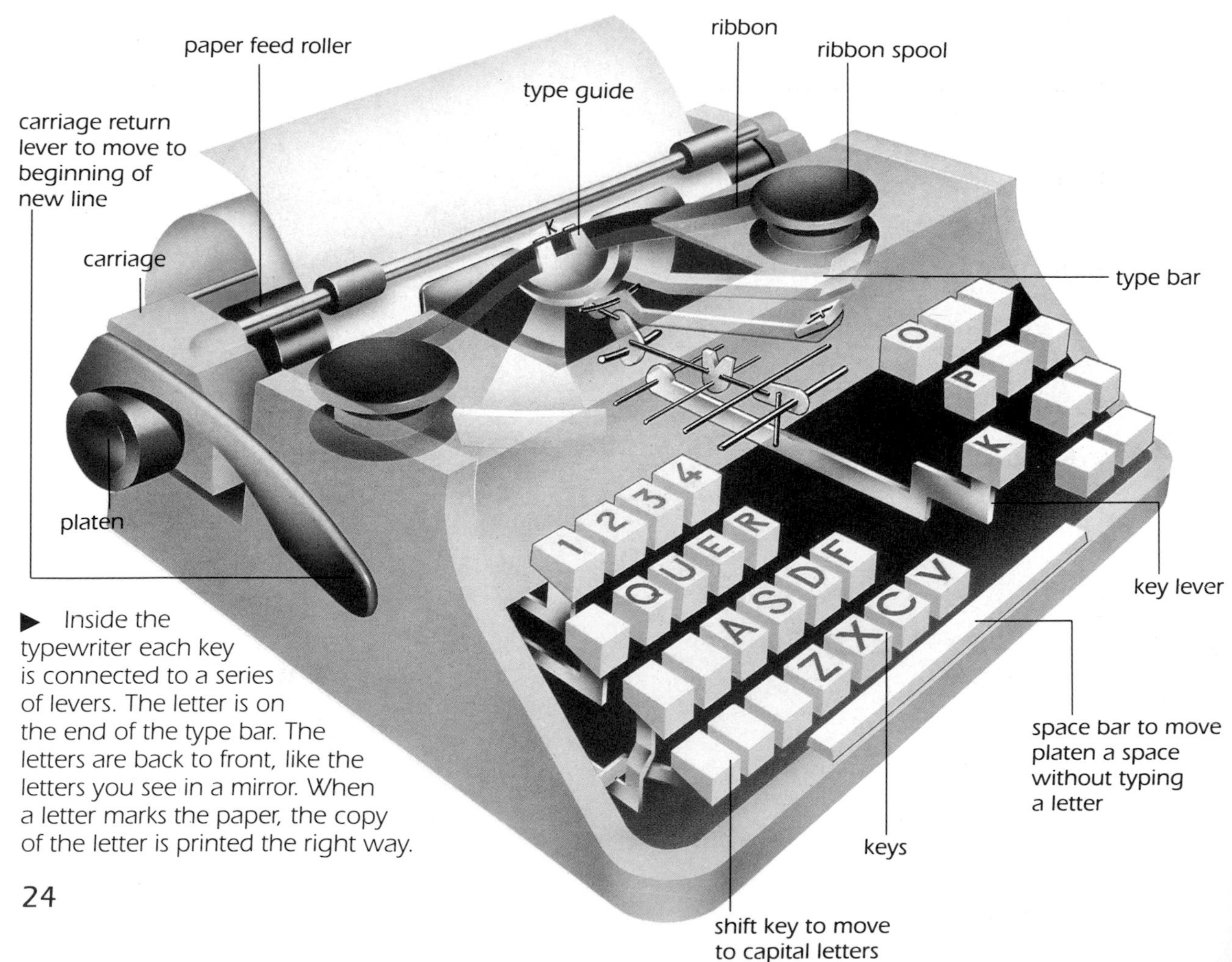

▶ Inside the typewriter each key is connected to a series of levers. The letter is on the end of the type bar. The letters are back to front, like the letters you see in a mirror. When a letter marks the paper, the copy of the letter is printed the right way.

Electric typewriters

Electric typewriters have an electric motor. The typist needs only to touch the keys lightly. Touching a key sends an electric signal to the motor. It moves the type bar to make a mark on the paper.

Instead of type bars, some electric typewriters have letters on the surface of a round ball. The ball moves around to put the letters in the right place as the keys are pressed. Now, most **electronic** typewriters have letters fixed to the ends of stalks joined onto a disc, rather like the petals of a daisy. These are called **daisy wheels**.

Computers and word processors

Computers have taken the place of typewriters in many offices. A computer has a keyboard like a typewriter. Instead of typing the letters directly onto paper, the computer saves the letters. One type of computer, called a **word processor**, does this. It stores the letters on a **magnetic disc**. The typist reads the words on a screen and corrects any mistakes before printing them.

The computer can make a copy of the text whenever it is needed. The word processor sends the letters to a computer printer. Some printers have daisy wheels like an electronic typewriter.

Daisy wheels

A daisy wheel is used in electric typewriters and some computer printers. Touching a key makes the wheel turn rapidly until the right letter is facing the paper. Then the letter strikes the paper. The carriage does not move along. Instead, the wheel moves along to the next space. At the end of the line the motor moves the paper up. Typing is much faster with a daisy wheel. It is also possible to change the wheel to use a different style of letters.

▶ The letters on the keyboard of a typewriter are not in alphabetical order. The layout was chosen so that the letters we use most often are not next to each other. This prevents the type bars from jamming together. The layout is the same on most computer keyboards. There are no type bars to jam but people are used to having the keys in this order.

Bicycles

A typewriter strikes a letter by using up and down motion. Up and down motion can be changed into rotary motion. This is done by the use of a **crankshaft**. A crankshaft is a rod, or crank, which moves up and down but turns around as it does so.

Using pedals

When you work the pedals on a bicycle, you move your feet up and down to make the pedals go around. This is because each pedal is fixed to a crank. The movement of the crank turns a small wheel beside the pedal. This wheel has cogs all around its edge. This type of wheel is called a **sprocket**. An endless chain fits over the cogs. The chain also fits around the cogs on a smaller

Mountain bikes

Mountain bikes are specially built for riding over rough ground. They are also called all-terrain bikes. They have heavier, stronger frames than other bikes and heavier, thicker tires. This bicycle has derailleur gears. There are two sprockets by the pedal and five sprockets on the rear wheel. This means that there are ten gears on this bike (5x2). When the rider is cycling on level ground or downhill, the rear-wheel sprocket needs to be small for high speed. When the rider needs to climb hills, the sprocket needs to be large so that the rear wheel turns with more force but less speed. When the rider presses the gear levers, the chain is moved onto different sets of sprockets.

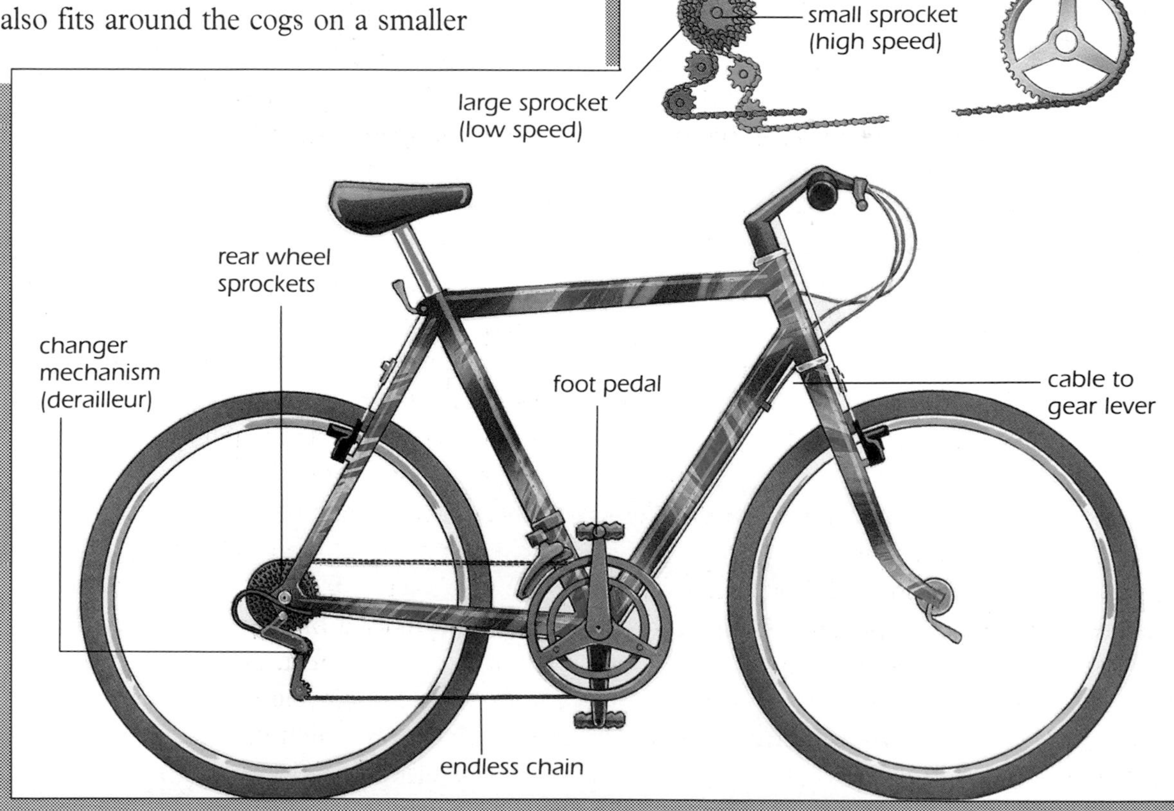

▲ Racing bikes are very light and have very slim wheels and tires. This makes them very fast to ride on smooth ground. These cyclists are taking part in a road race.

sprocket which turns the rear wheel of the bike. When you pedal, the sprocket beside the pedal turns and the chain moves around. The chain turns the sprocket beside the rear wheel which turns the wheel. The chain and sprockets send power from the cyclist's feet to the rear wheel.

Changing speed

There are many more cogs on a large sprocket than on a smaller sprocket. As the large sprocket turns around once, the smaller sprocket turns around a number of times. How many times depends on how many cogs there are on each sprocket. If the large sprocket has three times as many cogs, the smaller sprocket turns three times as fast. Using the difference in speed between small and larger wheels is called gearing.

Many bikes have sets of gear wheels called derailleur gears. These bikes have a number of different-sized sprockets. To change gears you move a small lever. This lifts the chain from one sprocket to another. This makes the rear wheel turn faster or slower. Changing gears makes it easier to cycle faster or uphill or against the wind.

Slowing down

If you stop pedaling, the bicycle slows down and then stops. This is because the tires rub against the ground. The ground acts like a brake. We say there is **friction** between the ground and the tires.

When you cycle, friction is always slowing you down. This is why a bicycle has narrow wheels. Only a small part of the bicycle is in contact with the ground. So there is less friction to slow you down. The only energy needed to power a bicycle comes from the movement of your feet. Nearly all the energy you put into the bicycle is used to make it move. Machines which use energy well are said to be efficient. The bicycle is the world's most efficient vehicle.

A gasoline engine

Like the bicycle, the gasoline engine uses a crankshaft to turn up and down motion into rotary motion. The first engines used up and down motion to pump water out of mines. These engines were powered by steam. About 200 years ago James Watt invented the crankshaft so that engines could be used to turn wheels.

People used steam to power slow-moving vehicles or "horseless carriages" on roads. Steam was noisy and needed a very heavy, bulky engine. So people tried to find ways of using other **fuels**. About 100 years ago, a German named Gottlieb Daimler invented an engine powered by **gasoline**. This was a light oil which burned easily when mixed with air and **compressed** in a small space. An engine powered by gasoline could be made small enough to be easily attached to a vehicle.

▼ *Each piston in a four-stroke engine goes through a continuous series of four strokes.*

The four-stroke cycle

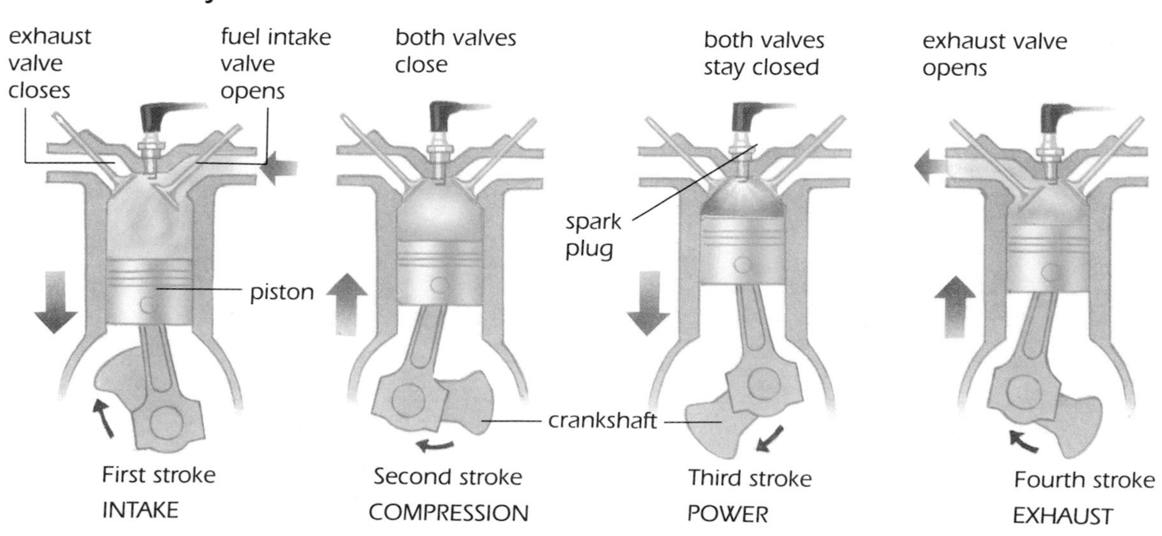

Inside the gasoline engine

Burning gasoline produces the power needed to turn the wheels of a car. Most gasoline engines contain round chambers called cylinders. Inside each cylinder is a round **piston** connected to a rod. The piston moves up and down to turn the crankshaft. Each movement, or stroke, plays a part in burning the fuel. The piston has to move up and down twice to complete the process. So we call this type of engine a **four-stroke engine**.

On the first stroke, the piston moves down. A gap, called a **valve**, opens up and a mixture of gasoline and air flows into the cylinder. On the second stroke, the piston moves up and the valve closes. The piston forces the gasoline and air into a small space at the top of the cylinder. An electric spark from the **spark plug** sets the fuel on fire. The burning fuel **explodes** and pushes the piston back down the cylinder. This push does the work of turning the crankshaft. So the third stroke is called the power stroke.

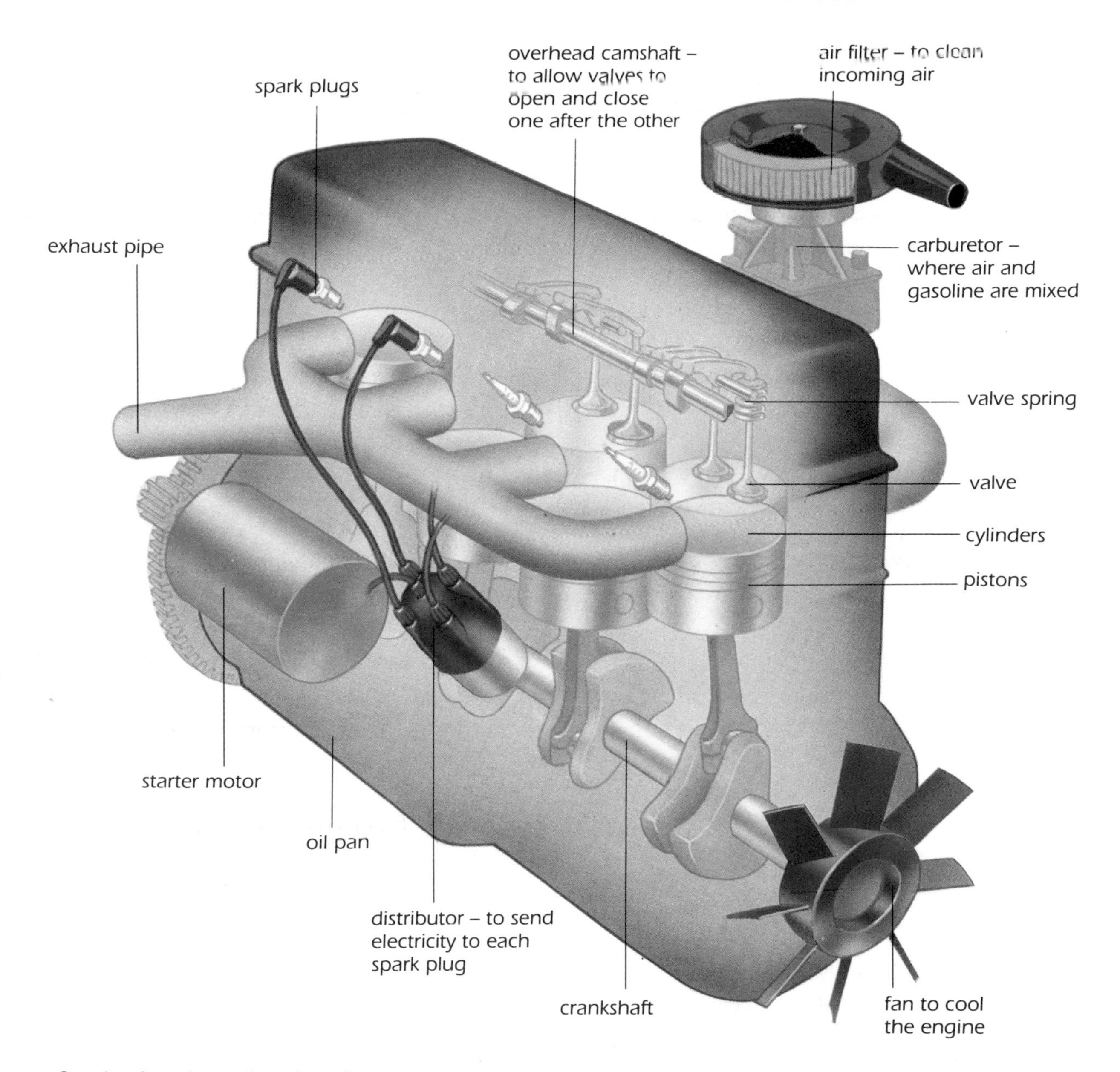

▲ *Most car engines have four, six, or eight pistons. They all go through the same series of four strokes but not all at the same time.*

On the fourth stroke, the piston moves up the cylinder again. As it does so, another valve opens at the top. The rising piston forces the gases from the explosion through this valve into the **exhaust pipe**. Then the piston begins the first stroke once more.

Internal combustion

The gasoline engine is an example of an internal combustion engine. The burning of the fuel (combustion) takes place inside the engine. The engine gets very hot and has to be made of strong metal so that it does not warp or bend. Oil is put into the engine so that the moving parts do not rub together. The parts could not move smoothly without the oil. They could cause a fire and would certainly wear out more quickly.

An automobile

You can see the main parts of an automobile in the picture. Power from the pistons turns the crankshaft. The rotary motion of the crankshaft passes through rods and gears to the wheels.

The driver can make the wheels turn faster using the **accelerator** pedal. Pressing the pedal sends the mixture of gasoline and air to the cylinders of the engine. When the driver presses the pedal down farther, more fuel goes to the cylinders. This makes the pistons move faster. As the wheels go faster, the car accelerates.

▼ This picture shows the main parts of an automobile. The engine produces the power. The power is transferred to the wheels by the use of the gears. The suspension system uses springs to keep the wheels on the road and to absorb the bumps and jolts.

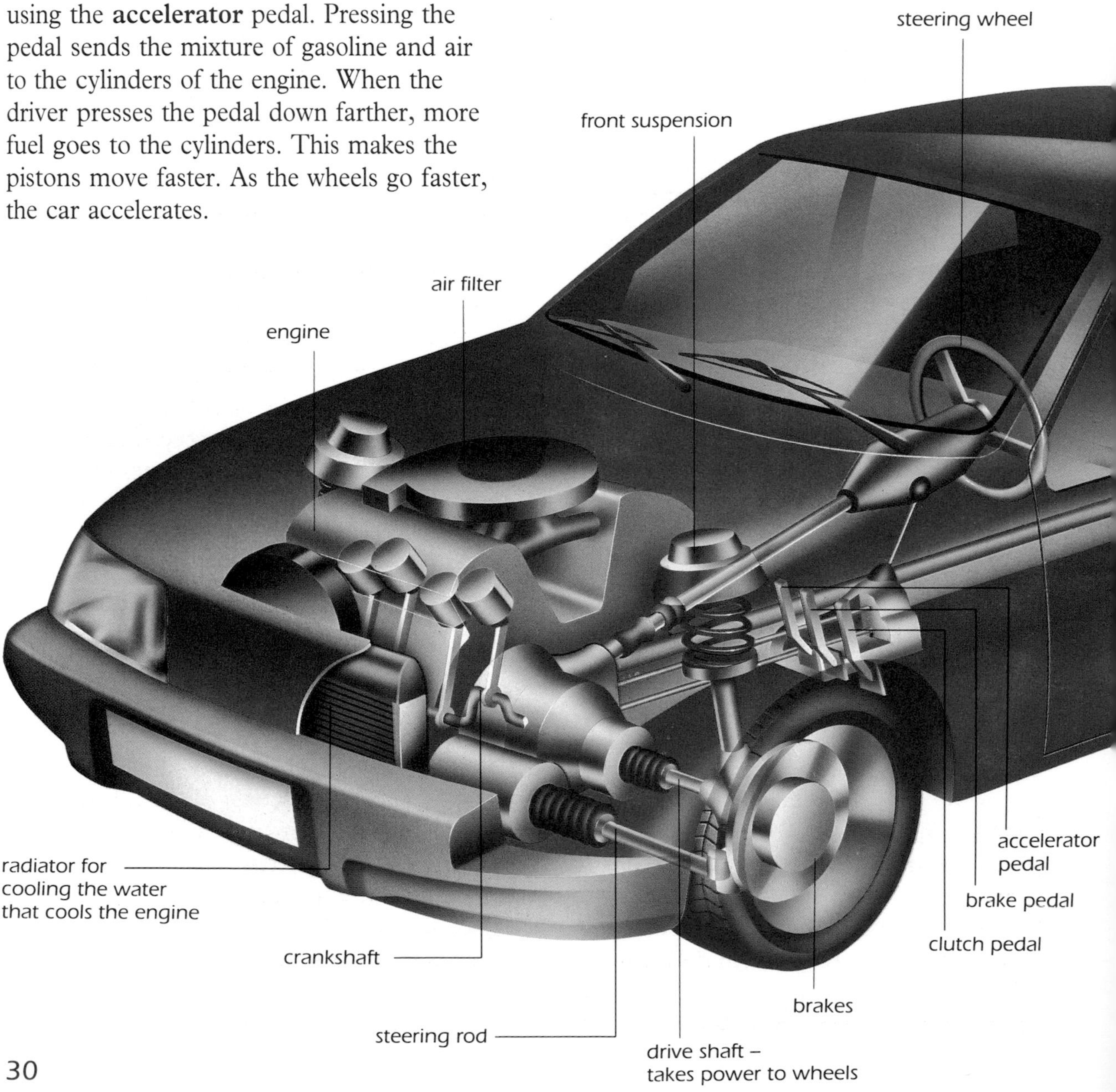

steering wheel
front suspension
air filter
engine
accelerator pedal
brake pedal
clutch pedal
radiator for cooling the water that cools the engine
crankshaft
brakes
steering rod
drive shaft – takes power to wheels

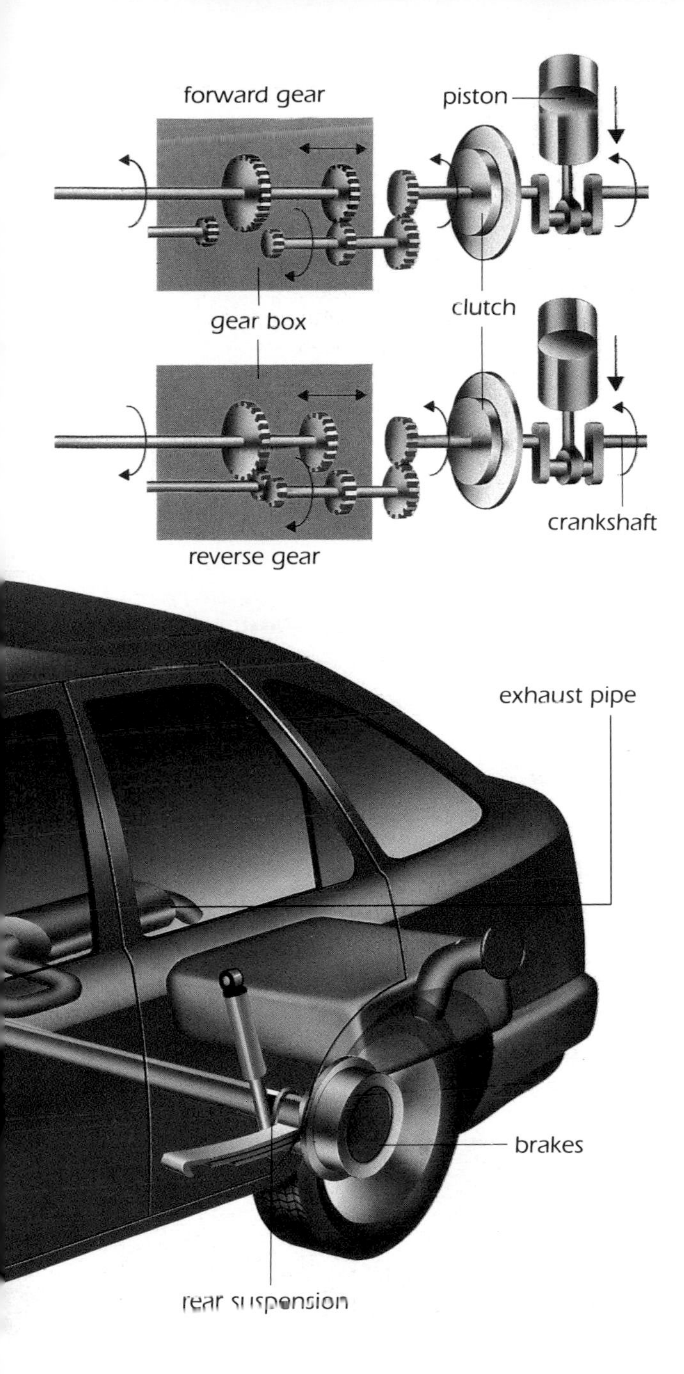

Using the gears

When you drive a car, you use gears to control the power of the engine. Most cars have five or six sets of gears. The different gears make the wheels turn in different ways.

When you want a car to move backward, you put it into reverse gear. When you want the car to move forward, you use the other gears. First or low gear makes the wheels turn slowly, with great force. You use first gear to start the car moving. As you move up a gear, the car moves faster. In top gear, the wheels turn quickly with little force. You use top gear to travel fast along flat roads. A car's top gear is usually fourth or fifth gear.

You change gears by pressing down on the **clutch** pedal. This breaks the link between the crankshaft and the gears. You then move the gearstick to the new gear. When you take your foot off the clutch pedal slowly, you link the new gear to the engine. Many cars change gear automatically when the driver presses the accelerator pedal. Cars like this have **automatic transmissions**.

Turning and stopping

The driver turns a steering wheel to change the direction of the car's motion. The steering wheel is linked by rods to the front wheels of the car. The driver must also be able to stop the car quickly. Pressing the brake pedal sends fluid down tubes to the brakes of all four wheels. Pressing on the fluid makes a very strong force. The fluid is able to move a brake pad against each wheel. The same amount of fluid goes to each brake. So the brakes all work together to stop the car smoothly. Brakes like these are called **hydraulic brakes**.

In comfort

A car does not bounce up and down too much even on bumpy roads. This is because the car rests on material which absorbs the bounces. The material forms the **suspension system**. It may be springs, rubber, liquids, or even air. The suspension system makes sure that the wheels stay on the road and that the car is always comfortable to ride in.

Motorcycles

The four-stroke engine is used for almost all makes of automobile. However, a large gasoline engine is not needed for every motor vehicle. Lawn mowers and motorcycles are not as heavy as cars. They do not need as much power to keep them moving, so smaller **two-stroke engines** are strong enough. The engine works in a similar way to a four-stroke engine. The main difference is that every second stroke gives power.

Inside a two-stroke engine

When the first stroke starts, there is gasoline and air at the top of the cylinder. As the piston moves up, it uncovers a gap, called a port. Gasoline and air are sucked into the bottom of the cylinder through this port. At the same time the piston squeezes the fuel at the top of the cylinder into a small space. The fuel is set on fire by a spark from the spark plug. The explosion forces the piston back down the cylinder. This is the second stroke, which does the work of turning the

▼ It is much easier to see the working parts on a motorcycle than on a car. The throttle is used like the accelerator pedal in a car. It makes the bike go faster. The throttle and the brake lever are usually on the handlebars where the rider can have more control of them.

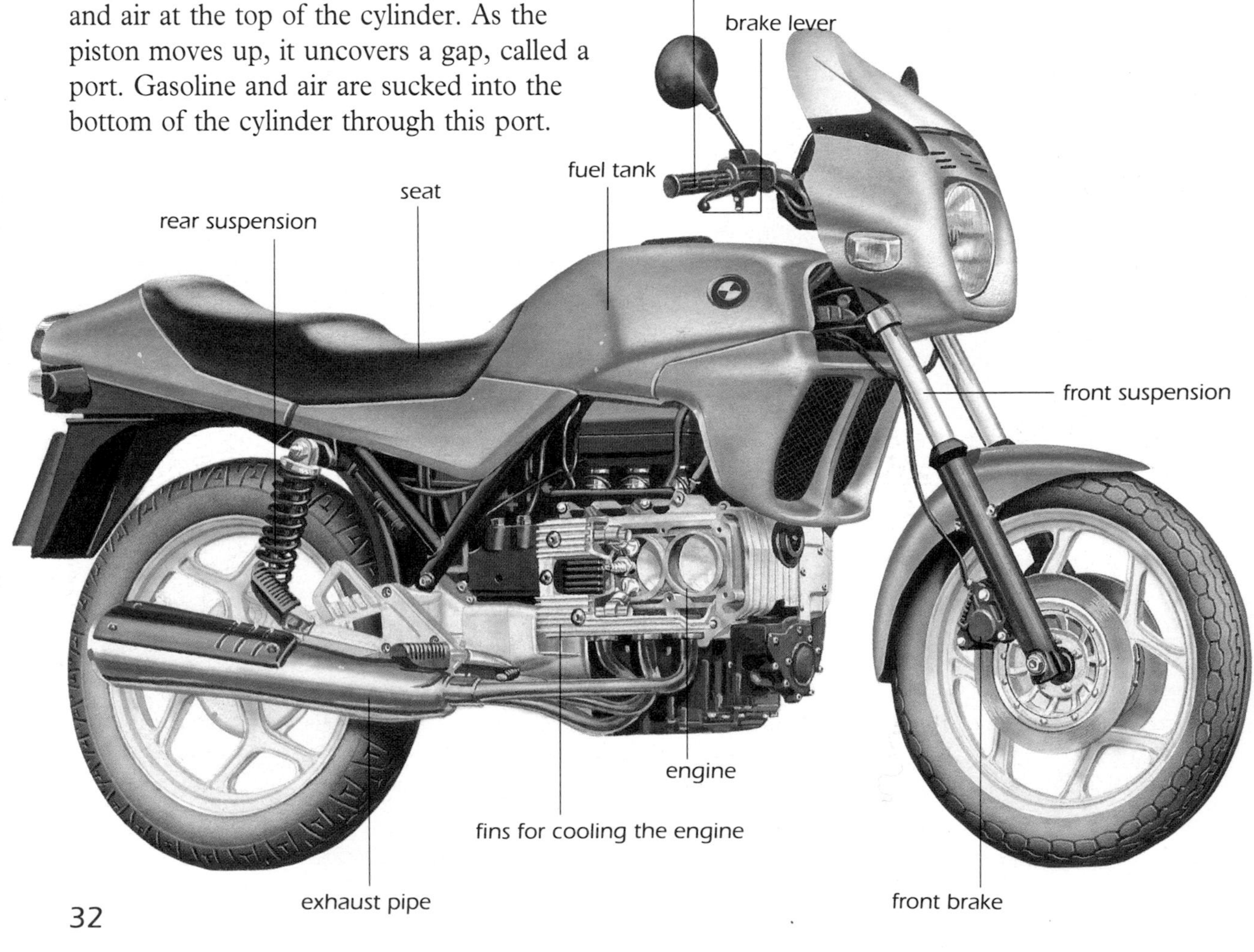

Suzuki 125 x Trakbike

Honda City Express Moped

Honda SS600

▲ There are many different types of motorcycles. Some look like bicycles with small engines to make the wheels go around. Other motorcycles are as powerful as cars.

wheels. As the piston moves down, it opens the exhaust hole at the side of the cylinder. The gasoline and air at the bottom of the cylinder rise upward. Then the gases created by the explosion rush out into the exhaust pipe. Then the piston moves up again and new gasoline and air are sucked in. The engine can be made to go faster by using the throttle, which does the same job as an accelerator on a car.

Keeping the engine cool

When the gasoline burns in the cylinders, it reaches temperatures of over 1292°F (700°C). Less than a quarter of this heat is used as power. A cooling system has to be used to stop the heat from damaging the engine. Most four-stroke engines are cooled by water passing around the engine. Air cooling is almost always used in motorcycle engines and in some cars because it is less bulky. The cylinders have metal fins all around them. The heat is spread out evenly through these fins. Cold air is forced over the fins by a fan.

Engine power

Not all engines have the same number of cylinders. Cars have engines with at least four cylinders. Many motorcycles have only one cylinders. Many motorcycles have only one but others may have two, three, or four. A four-cylinder engine is more powerful than a one or two-cylinder engine. There are a few very powerful motorcycles which have six cylinders.

A power drill

Rotary motion can be used to do many jobs. A power drill uses rotary motion to drill holes. An electric motor provides power for the drill, so it can be used anywhere in the home near an electric socket.

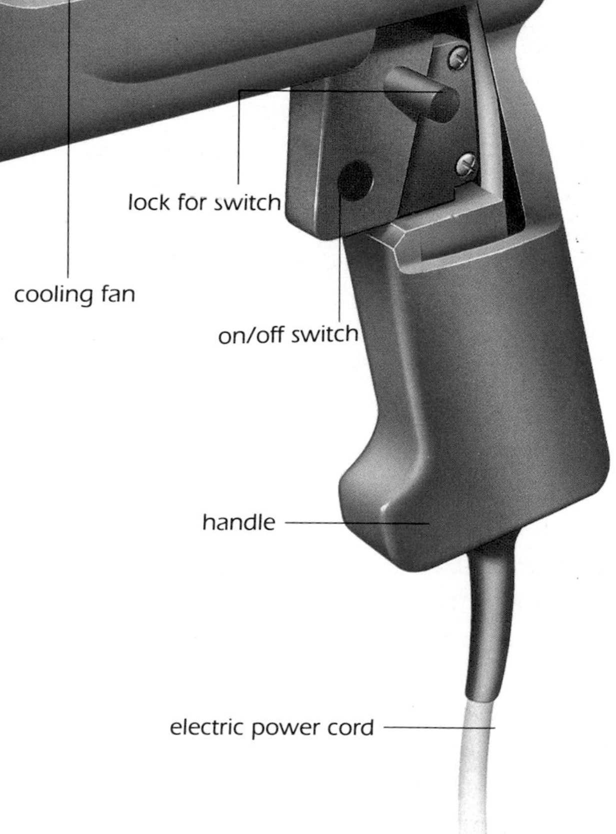

▲ *A power drill can bore through harder materials than a hand drill. The power drill is also quicker, less tiring to use, and more accurate.*

Using a power drill

At the end of the drill are strong jaws called the **chuck.** This grips the sharp-pointed metal **drill bit** tightly. The motor turns the drill bit around at high speeds. When pressed against a wall, the drill bit pierces the surface and soon bores a hole.

We can put different size drill bits in the chuck on a power drill. The larger-sized drill bits make bigger holes. The power drill can bore holes in many materials, such as brick, wood, and metal. Very tough steel drill bits are used to bore through the toughest materials.

Roadworks

This drill uses an up and down motion to split the road surface. It is called a **pneumatic drill** because it gets its power from compressed air. Air squeezed into a small space has great force. This compressed air pushes a sharp metal blade into the surface of the road. The air enters the pneumatic drill in short bursts. So the blade hits the road in a series of rapid blows.

◀ Power drills can be used to do a variety of jobs. This woman is using a wire brush attached to a drill to clean rust off a car.

We can use the rotary motion of the power drill to do other jobs as well. We can smooth the surface of a piece of wood by attaching a sander to the drill. The sander is a metal disc covered with sandpaper. The disc rotates, rubbing the surface much faster than you could rub it by hand. Many other tools can be fitted to a drill, such as a wire brush and a polisher. A power drill can turn a circular saw to cut wood. It can also turn a grindstone to sharpen knives.

Drilling through rock

Tunnel workers use drills to bore small holes in hard rock. The workers fill the holes with **explosives**. These blow out the rock to leave a tunnel-sized hole. Oil workers may need to bore through hundreds of feet of rock and stone to find oil. These workers use drills on the ends of long pipes. Again, these drills use rotary motion.

A dentist's drill

Dentists use a small power drill before filling a tooth. The drill bores away the bad part of the tooth. The drilling causes friction, which makes the tooth hot. This is why the dentist uses a spray of water to cool the tooth when drilling.

Mixing and cutting

Many years ago, people working in a kitchen did all the jobs by hand. Cooks mixed ingredients like flour, sugar, milk, and eggs with a wooden spoon or with a hand whisk. They chopped fruit, vegetables, and meat with a knife. Today, cooks in restaurants use tools similar to the power drill to do these jobs. Some people have smaller versions of these machines to use at home.

Mixing food

The electric mixer can blend ingredients much quicker and with far less effort than a hand mixer. An electric motor turns the beaters around. The beaters mix the food.

An electric food mixer can also be used to beat food. Like the power drill, this machine can also do other jobs. The food mixer has a wide range of tools, or attachments. The cook fixes the attachments to different parts of the mixer. The rotary motion of the mixer turns the attachments.

Pressing or kneading dough to make bread is hard work. A hook-shaped attachment called the dough hook can knead dough. The potato peeler attachment removes the skin of a potato in seconds. The coffee mill grinds coffee beans to make freshly ground coffee. Another attachment squeezes juice out of oranges or lemons.

A **blender** changes solid food into a liquid called purée. Cooks use blenders for fruits and vegetables, such as carrots and tomatoes. The purée is used to make soups, sauces, drinks, and baby food.

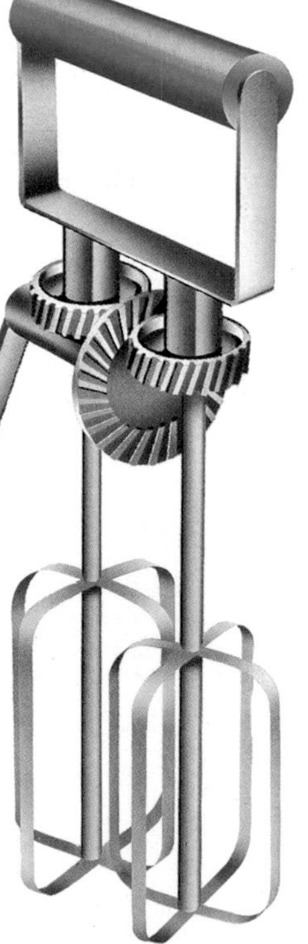

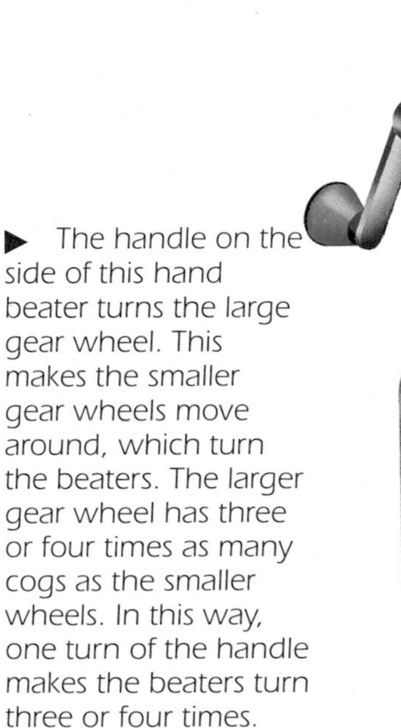

▶ The handle on the side of this hand beater turns the large gear wheel. This makes the smaller gear wheels move around, which turn the beaters. The larger gear wheel has three or four times as many cogs as the smaller wheels. In this way, one turn of the handle makes the beaters turn three or four times.

Chopping food

Food processors can do a number of jobs which cannot be done with a mixer. There is an electric motor in the heavy base of the processor. A metal blade fits into the middle of the processor, above the motor. The motor turns the blade around. It can rotate at fast or slow speeds depending on the job. The blade chops, grinds, or minces the food.

The cook can take out the blade and fit another attachment in its place. The shredding disc spins around to shred vegetables into long or short lengths for salads. The slicing disc slices food, such as fruit and vegetables. A plastic blade can be used to mix ingredients together. The food processor can also squeeze the juice out of fruit, make purée, and knead dough.

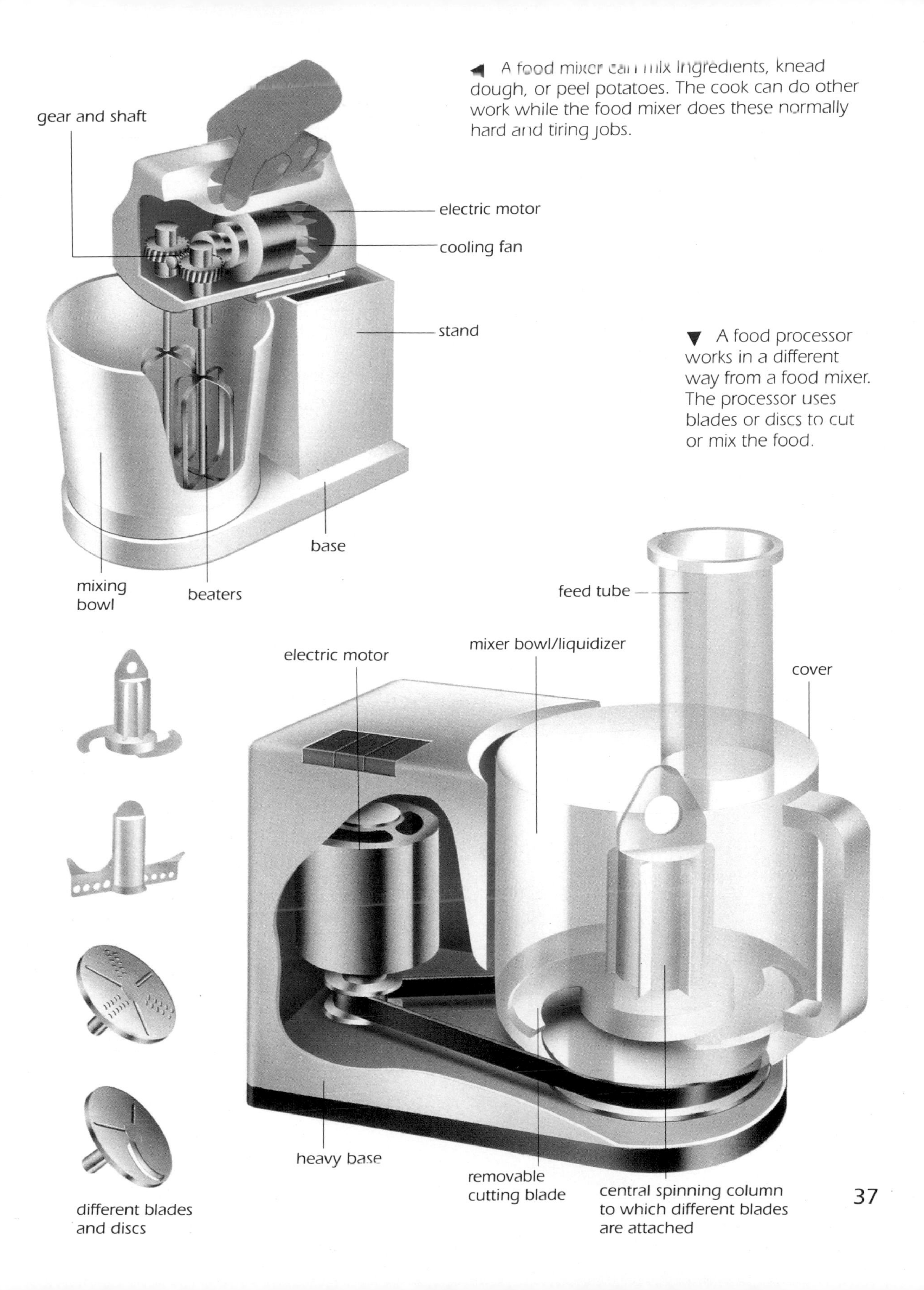

◀ A food mixer can mix ingredients, knead dough, or peel potatoes. The cook can do other work while the food mixer does these normally hard and tiring jobs.

▼ A food processor works in a different way from a food mixer. The processor uses blades or discs to cut or mix the food.

A combine harvester

For thousands of years, farmers all over the world have used sickles to cut their wheat and other crops. A sickle is a curved blade with a sharp inside edge. The farmer swings the sickle sideways to slice the wheat.

The next stage of harvesting is to separate the wheat from the stalks. Farmers have to beat the wheat with sticks. This is called **threshing** the wheat. The farmers then throw the ears of wheat in the air. The wind blows away the loose outer case, or chaff, leaving behind the grains.

These methods of harvesting by hand are still used in many countries. But we now also have machines which can do a lot of the hard work.

How the combine harvester works

The combine harvester does all the jobs involved in harvesting wheat in a few seconds. It is like a small factory on wheels.

The dividers at the front of the combine harvester guide the wheat into the machine. Sharp blades are fixed behind the dividers. These blades lie across the combine just above the ground. The blades move rapidly from side to side to cut the stalks of wheat close to the ground. A long screw, called an

▶ The combine harvester combines the actions of cutting and threshing the wheat.

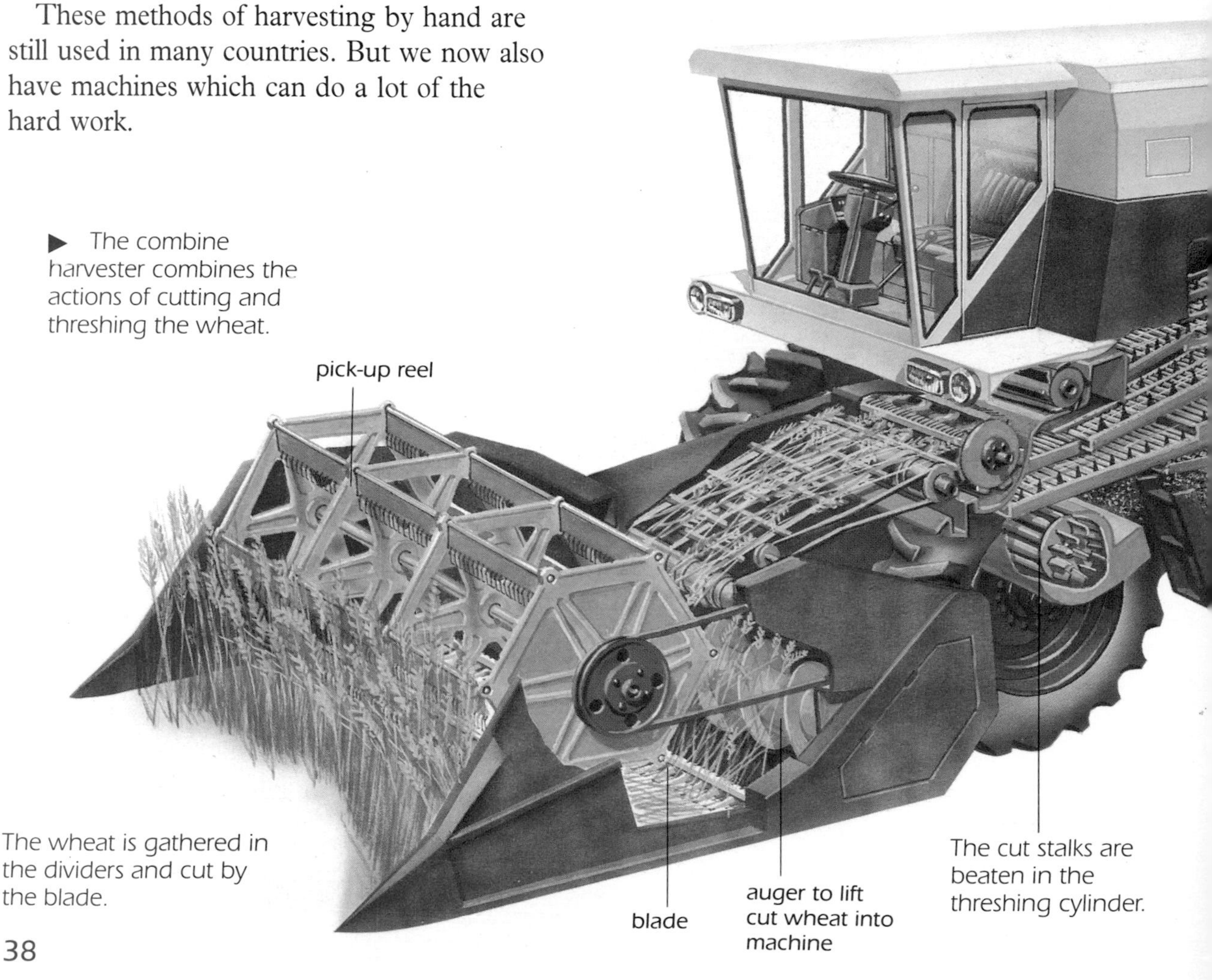

pick-up reel

The wheat is gathered in the dividers and cut by the blade.

blade

auger to lift cut wheat into machine

The cut stalks are beaten in the threshing cylinder.

auger, turns around to carry the cut wheat up into the threshing cylinder. This rubs the wheat to separate the ears from the stalks, called straw. Straw cannot be eaten, but is used to line the sheds where animals live. The straw moves along conveyor belts called straw walkers, and falls off the back of the machine. Later, the straw is packed tightly into bales and stacked on the farm.

straw walkers

auger for unloading grain

chaff blown out here

The straw is carried along the straw walkers to the back of the machine where it falls off.

The grain and chaff fall onto the sieves and the chaff is blown away.

Cleaning the wheat

The ears of wheat fall through the rack of the straw walker. The wheat still has a lot of straw and dirt mixed with it. The combine shakes the ears of wheat from side to side across sieves. At the same time, fans blow air through the wheat. The chaff is blown to the back of the harvester. The heavier grains of wheat are left behind. The wheat is stored in a large tank. When the tank is full, the combine driver unloads the wheat. A tube swings out from the combine. It has an auger inside which moves the ears of wheat up. The wheat pours out of the tube into a truck driven alongside the combine.

▲ Tractors with trailers can drive alongside combines to collect the wheat. The wheat is sprayed into the trailer from an auger.

Mowing

We do not need a combine harvester to cut grass but it is still hard work to do by hand. When grass is very long we have to use a sickle or hand shears which are like large scissors. But there are now several types of machines we can use to cut shorter grass.

Lawn mowers can be powered by a person pushing them. Some types get their power from an electric motor or a small gasoline engine. Electric lawn mowers get their power through cables carrying electricity from the house. Motor mowers have two-stroke gasoline engines, like those on a motorcycle. There are two main types of mower: the cylinder mower and the rotary mower.

The cylinder mower

Cylinder mowers have several blades fixed together in the shape of a cylinder. The blades run the length of the cylinder. They are slightly twisted to make the cylinder stronger. The sharp edge of the blade is on the outside edge. The cylinder is fixed across the mower with the blades along the ground. Behind the cylinder is a curved plate with a fixed blade at the bottom. As the cylinder

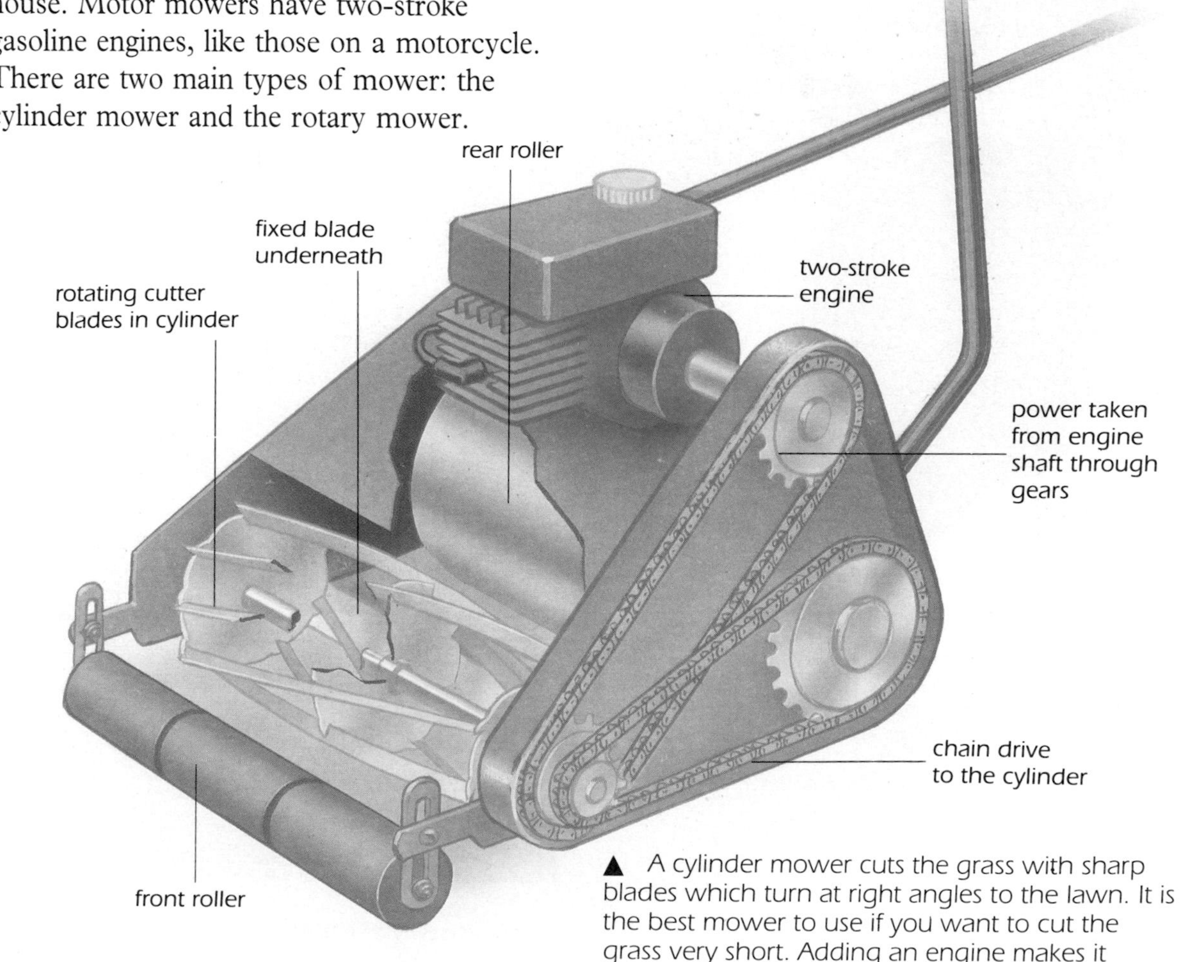

▲ A cylinder mower cuts the grass with sharp blades which turn at right angles to the lawn. It is the best mower to use if you want to cut the grass very short. Adding an engine makes it much easier to use.

A hover mower

The hover mower is a special kind of rotary mower. It floats like a **hovercraft**. A fan sucks in air at the top of the mower and blows the air out at the bottom. The mower floats on this cushion of air. This makes the mower very easy to push.

turns, the moving blades meet the fixed blade. Any grass trapped between them is cut. The gardener can raise the blades to cut long grass or lower them to cut short grass.

Behind the fixed blade is a large, solid cylinder, or roller, which acts as a wheel. The roller is joined to the cutting cylinder by an endless chain. As the mower is pushed the roller turns and drives the blades. Some mowers are pushed by a person but many have a motor which drives the endless chain. Some large lawn mowers have seats for gardeners to sit on. Cylinder lawn mowers like these are used to cut large areas of grass, such as in a park or golf course.

The rotary mower

The rotary mower has one long blade underneath the motor. The blade rotates in a different direction from those in a cylinder mower. The blade turns across the ground, like a bicycle wheel on its side. As the blade spins round, it cuts the grass. Again, the gardener can raise or lower the blade.

Many rotary mowers move along the lawn on four wheels. Others have two wheels at the front and a roller behind which flattens the grass.

▼ The blade of a rotary mower spins rapidly just above the ground. This mower is good for cutting rough grass and long grass.

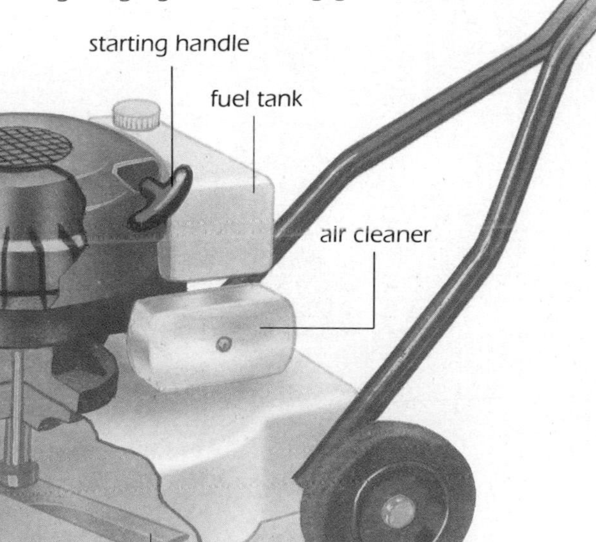

Moving pictures

A film **projector** uses several types of motion at the same time. This machine fools people into thinking they are seeing moving pictures. The first movie audiences ducked in their seats when they saw a train coming toward them on the screen! The audience was actually watching a number of still pictures, or frames, one after the other.

The pictures are on a long strip of film wound onto a reel. Ninety years ago, the film was moved through the projector by someone turning a handle. In those days there was always a slight flicker between pictures when people watched a film. This is why people used to call movies the flicks.

How a movie projector works

Today, an electric motor provides the power for the film projector. The beginning of the reel of film is wound around sprockets in the projector. The film has a line of holes on either side. These fit into the small teeth on the sprockets. The teeth hold the film firmly in place. The electric motor makes the

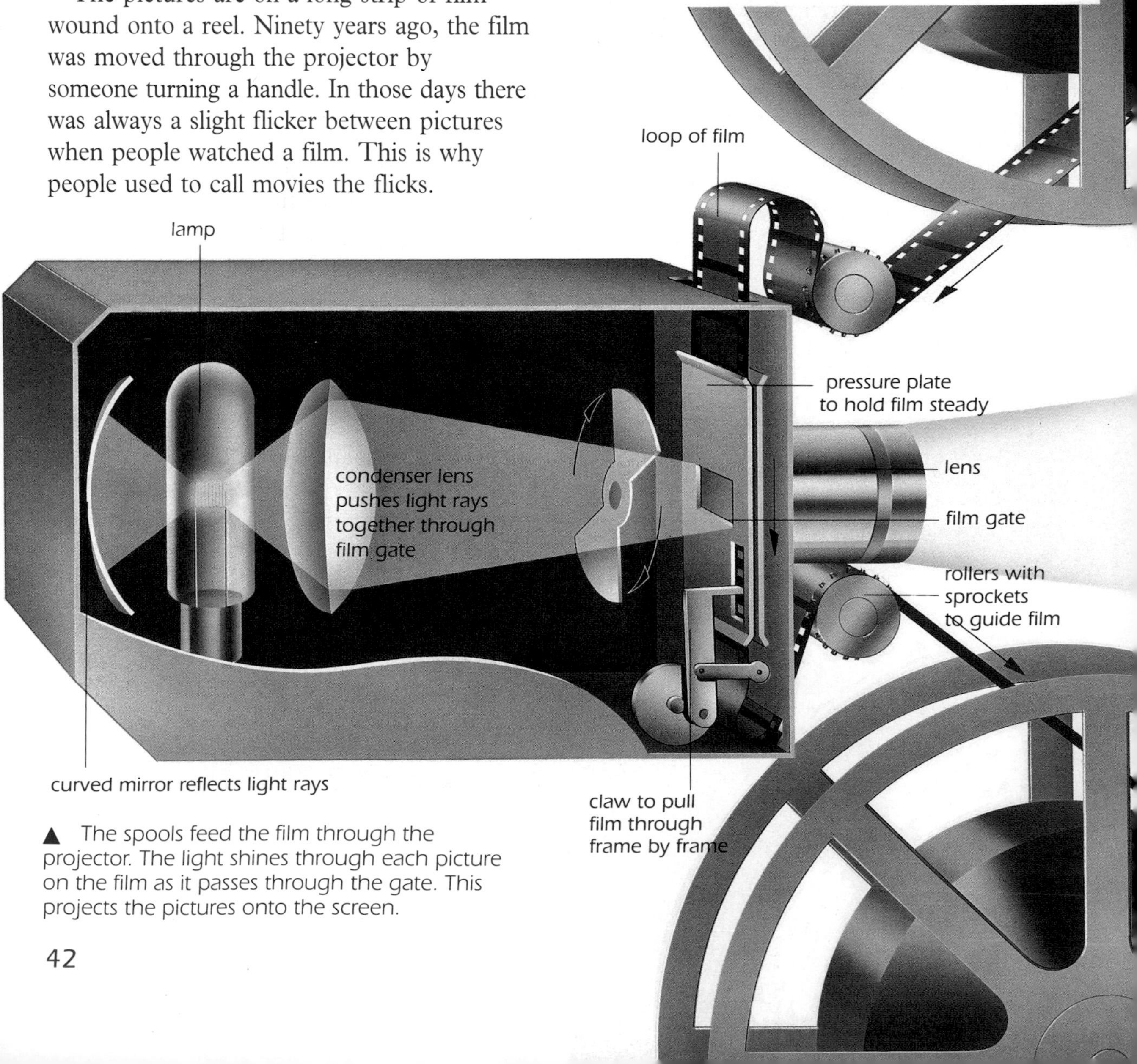

▲ The spools feed the film through the projector. The light shines through each picture on the film as it passes through the gate. This projects the pictures onto the screen.

sprockets turn around. The sprockets pull the film through an opening in the projector called the **film gate**. Light from a powerful lamp shines through the picture in the gate to the screen. This projects the picture onto the screen. The film stops moving for a tiny part of a second. Then the next picture is pulled down to take its place. So the projector is stopping and starting all the time.

Fooling the eyes

If you could see the film move, the picture on the screen would look blurred. So the projector contains a piece of metal called a shutter. Each time the film moves, the shutter cuts off light to the screen. The cinema is in total darkness while the picture changes. The shutter and the film move so quickly that the first picture stays in your eyes until the next picture appears an instant later. Your eyes and brain do not notice the darkness. Instead, your brain thinks all the movement is coming from the picture on the screen!

film spool or reel

▶ Motion-picture film is made up of thousands of tiny pictures. Each picture is slightly different from the one in front. The pictures are shown quickly on the screen, one after the other, so we cannot see separate pictures. Instead, we see movement on the screen.

takeup spool

Film projection

Film moves through a projector at a rate of about 24 frames a second. But the projector shows each frame twice, so we see 48 pictures each second. This reduces flickering.

Did you know?

Slow motion

Some motion is so slow it cannot be seen. A machine has been invented that can measure movements as slow as one trillionth of a millimeter per minute (a millimeter is about one twenty-fifth of an inch) or 1 yard in two billion years.

Spinning around

The earth orbits the sun at a speed of over 62,000 mph (100,000 kph). This is over 17 miles (25 kilometers) every second.

Machine motion

Robots are run by computers. They can be programmed to make many different movements. Robots are often used in factories where the same movement has to be repeated again and again.

Speed records

In 80 years the gasoline-driven car has developed from a speed of 3 mph (5 kph) to a maximum speed of 417 mph (673.5 kph). The first successful gas-driven automobile was made by Karl Benz in 1885. It went at about 3–4 mph (5–6 kph). The speed record is 417 mph (673.5 kph), which was reached in Utah on November 12, 1965.

Fun rides

People use motion for amusement. There are many parks where machines spin people around or push them high into the air. The cars on this giant ride are driven up to the top by a motor and an endless chain. The weight of the car brings it back down again very quickly.

Fast motion

The fastest moving object ever on earth was a tiny plastic disc. It was fired by a laser in Washington, DC, in August 1980. The disc reached a speed of just over half a million kilometers an hour.

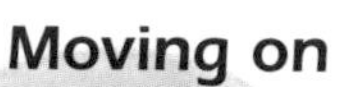

Moving on

In the future people will find other ways of using motion to help us in our everyday lives. In the last 100 years or so, people have invented cars, motorcycles, escalators, elevators, combine harvesters, aircraft, hovercraft, and hydrofoils. Who knows what new forms of moving machines people will invent in the next 100 years?

Skimming the surface

A **hydrofoil** skims quickly over a lake or sea. This is because most of the boat travels through air not water. When an object travels on land or on water, it rubs against the surface. This rubbing (friction) slows it down. Air slows an object down much less than water.

Glossary

accelerator: a pedal in a vehicle. Pressing the accelerator pedal sends more fuel to the engine and makes the vehicle go faster.
anchor: a device with arms with a hook on the end. The hooks are used for grabbing and holding.
auger: a device like a screw or corkscrew. The twisting action of large augers can be used to move liquids or small objects to a higher level.
automatic transmission: a system in a car which changes the gears. The gear changes are made when certain speeds are reached without the driver having to take any action.
automatically: moving or working on its own.
balance: to make two things equal in weight. On scales, equal weights make the balancing bar level.
blender: a tool which makes a liquid out of a solid food, such as fruits or vegetables.
bridge crane: a crane where the lifting mechanism hangs from a beam across the middle.
cage: a lift which takes miners up and down a mine shaft.
carriage: the part of a typewriter which carries the paper and platen.
chuck: the part of a power drill which tightens to grip the drill bit or any other attachment fitted to the drill.
clutch: a device in an automobile which allows the driver to disconnect the engine from the parts which make the car move, especially while changing gears. The clutch is released by pressing the clutch pedal.
cog: a tooth on the edge of a wheel.
compressed: squeezed into a small space.
conveyor belt: belt of strong material which is moved over rollers by a motor. Objects put onto the belt are carried along.
coordination: when movements or actions are arranged so that they all work together.
counterweight: a weight used to press down on one end of a rope or pole when an object is being lifted at the other end.
crankshaft: a shaft which turns up and down motion into rotary motion. One end of the shaft is joined to a part that moves up and down and the other end to a part which moves around.
daisy wheel: a device for holding letters in an electronic typewriter. The wheel has letters fixed to the ends of stalks. The other ends of the stalks meet at the center of the wheel.
damper: a device which stops vibrating piano strings.
drill bit: a sharp metal tool used to bore holes. It can be fixed in the end of a power drill.
effort: the force used to do a job or work, such as lifting a heavy object.
electric motor: a machine which gets its energy from electricity and uses it to make other machines work.
electronic: any device which works by means of electrical signals switching on and off.
endless chain: a belt, cable, or chain, which does not have an end or a beginning, such as a bicycle chain.
energy: the power to do work. People get energy from food. Engines get energy from fuel.
escalator: moving steps which go around and around on an endless chain.
exhaust pipe: a pipe which takes away the gases made when a fuel burns.
explode: get much bigger very quickly. Anything which explodes pushes with great force.
explosives: substances which will explode.
film gate: the opening in a film projector. A powerful lamp sends a beam of light through the gate to project a picture onto a screen.
first-order lever: a simple lever, such as a seesaw, with the fulcrum between the effort and the load.
food processor: a kitchen machine which processes food by cutting, slicing, chopping, or shredding it.
force: strength or energy which produces motion.
four-stroke engine: a gasoline engine in which the moving parts have a cycle of four stages or strokes.
friction: the rubbing of one surface against another. Friction slows machines down.
fuel: a substance which is burned to produce energy. Gasoline and coal are fuels.
fulcrum: the point at which a lever turns or balances.
gasoline: a liquid which is made from oil and burns easily.

gear wheels: a group of wheels with teeth around the edge used to change the speed of one wheel. The teeth fit between the teeth in other gear wheels. As each gear wheel moves around, it can turn many other gear wheels.
girder: a large beam made of wood, iron, or steel used as a support in building.
governor: a device which limits the speed of a motor or engine by controlling how much fuel is allowed to flow.
grab: an attachment to a crane with large jaws which open and shut. The grab scoops up loose materials, such as sand or soil.
hovercraft: a vehicle which travels on a cushion of air. The hovercraft can move over land or sea without touching the surface.
hydraulic brakes: brakes which work because of the effects of pressure on a fluid.
hydrofoil: a boat which is lifted up on winglike foils to skim over the water.
jib crane: a crane with a long lifting arm called the jib.
joints: where two moving parts meet. We move our body at the joints, such as the wrist or elbow.
kinetic energy: the energy in something when it moves.
lever: a tool, such as a shovel, which helps people move or lift things. When effort is applied to one part of the lever, it magnifies the effort and applies it to a load at another part.
load: the weight or force of an object moved by a lever or other machine.
magnetic disc: a flat, circular plate which can record electrical signals. The disk is coated with a magnetic material.
muscles: the parts of the body which produce motion. Muscles are made of bunches of fibers and are found all over the body.
pendulum: a weight which can swing freely. When pushed, the pendulum swings back and forth.
percussionist: the musician in an orchestra who plays the instruments which have to be struck such as cymbals, xylophone, and bells.
piston: a round piece which fits inside a cylinder and moves up and down on a rod.
platen: the rubber roller which holds the paper on a typewriter or computer printer.
pneumatic drill: a power drill which uses compressed air to drive a blade into a surface.
power: energy which can be used to do work.
production line: the system by which goods are put together in a factory as they pass along a conveyor belt.
projector: an instrument which uses light to throw an enlarged image on a surface.
pulley: a wheel with a groove around the rim for a rope or cable that is used to lift heavy objects.
robot: a machine that can be programmed to do jobs automatically.
rotary motion: movement which goes around and around, such as a wheel spinning.
scale: a machine used for weighing objects or people. They stand on a platform hanging from the short arm of a steel beam. A weight is moved along the long arm until the beam balances.
second-order lever: a lever with the fulcrum at one end and the effort at the other such as a wheelbarrow. The load moved by the lever is between the fulcrum and the effort.
shaft: a long narrow pole such as a crankshaft, or a tall, upright hole such as an elevator or well shaft.
spark plug: a device which produces a spark from an electric current. The spark sets fire to the gasoline in a gasoline engine.
sprocket: a wheel with teeth or cogs around the edge, which can be turned by a chain or belt.
steam engine: an engine in which steam is used to force the moving parts up and down or around.
stored energy: the energy stored in something, such as food or gasoline, that can be used when needed by the body or a machine.
suspension system: the springs, rubber, or other materials which stop a vehicle from bumping over a rough surface. The suspension system lies between the main body of the vehicle and the wheels.
third-order lever: a lever in which the fulcrum is at one end while the load is at the other end such as a pair of tongs. The effort is applied between the load and the fulcrum.
threshing: beating crops, such as wheat or rice, in order to make the grain fall off the stalks.
two-stroke engine: a gasoline engine in which the moving parts have a cycle of two stages or strokes.
type bar: the lever in a typewriter which has a letter, number, or sign fixed to one end. There is a type bar for each letter.
valve: a device that opens or closes to control the flow of a liquid or gas.
vibrations: rapid movements back and forth, such as those of a guitar string when you pluck it.
weighing scales: a machine used for weighing things, such as food or people.
word processor: a computer used to type letters and other written documents. A word processor can hold documents in its memory to be printed out at another time.

Index